Starry Night® College
Erin O'Connor
SANTA BARBARA CITY COLLEGE

Observation and Research Projects
Steve McMillan
DREXEL UNIVERSITY

ASTRONOMY TODAY

Chaisson
McMillan

SEVENTH EDITION

Addison-Wesley

Boston Columbus Indianapolis New York San Francisco Upper Saddle River
Amsterdam Cape Town Dubai London Madrid Milan Munich Paris Montreal Toronto
Delhi Mexico City São Paulo Sydney Hong Kong Seoul Singapore Taipei Tokyo

Publisher: Jim Smith
Executive Editor: Nancy Whilton
Editorial Manager: Laura Kenney
Senior Project Editor: Katie Conley
Managing Editor: Corinne Benson
Production Project Manager: Mary O'Connell
Production Service and Composition: PreMediaGlobal
Cover Design: Derek Bacchus
Manufacturing Buyer: Jeffrey Sargent
Text and Cover Printer: Edwards Brothers Malloy
Cover Image: Carina Nebula, NASA, ESA, and M.Livio and the Hubble 20th Anniversary Team
   (STScL)

Starry Night® and the **Starry Night** logo are trademarks of Simulation Curriculum, a Minnesota Corporation, and are used under license.

Many of the designations used by manufacturers and sellers to distinguish their products are claimed as trademarks. Where those designations appear in this book, and the publisher was aware of a trademark claim, the designations have been printed in initial caps or all caps.

ISBN-13: 978-0-321-75307-6
ISBN-10:   0-321-75307-0

Addison-Wesley
is an imprint of

www.pearsonhighered.com   2 3 4 5 6 7 8 9 10—V069—15

Starry Night® College Activities

Erin O'Connor

SANTA BARBARA CITY COLLEGE

Starry Night College Activities

Contents

Starry Night College Activities

(Erin O'Connor, Department of Earth & Planetary Sciences, Santa Barbara City College)

INTRODUCTION:

Welcome to the Starry Night College *Activities.* This series of simple activities is unique in that they have been designed to help "*you,*" the student, make a personal connection with "*your*" current evening sky. In general, we ask that you do not set the software to some arbitrary date, time, and location for the purpose of carrying out an activity. Every effort has been made to orient the material to be relevant from your perspective. This will help you connect what you are learning in your astronomy course with what you see in your night sky.

These are exploratory activities designed to guide you through the night sky with the use of the Starry Night College software. This software's exceptional qualities are its realistic look and feel, ease of use, and tremendous astronomical capabilities. We hope that these activities will serve as platforms from where you will launch into explorations of your own. You may do this at any point during an activity. Simply click the **back** button as needed to return to your previous configuration to carry on with the activity.

These activities are not designed as evaluation tools, but rather serve as self-guided tours of celestial phenomena or topics related to your lecture course. We do not introduce and/or teach significant amounts of new material. Your textbook, in conjunction with your course lectures, will present the material. Here we simply try to help you visualize celestial cycles, patterns, motions, and objects—and to bring focus to capabilities being demonstrated and/or displayed by the software.

Many of the activities encourage student interaction by having you answer simple questions that can be easily obtained as you work through the material. These questions may on occasion seem overly simplistic, but they help bring focus to key concepts and ideas that may be discussed in your astronomy course. The primary goal here is not the answer to the question, but rather the process of having you work through and interact with the software in getting to that answer. Once you know the process, you can use Starry Night College to seek out similar information for any planet, moon, star, or celestial object. We are really focusing on the process, not the specific data on specific items.

In many cases, similar questions with similar answers can be found in multiple activities. Since the activities are designed to stand alone and be independent of each other, overlap is unavoidable. Simply skim through these sections, and/or use the answers you obtained from a preceding activity should overlap occur. This allows your instructor the greatest flexibility in selecting activities for an effective program of study to complement your astronomy lecture course. Typically, only a handful of activities may be selected for a full semester's course. Many activities are redundantly similar in process, yet may focus on a different planet and/or area of the sky.

Many students live in metropolitan areas severely limited and/or restricted in viewing of the night sky. For this reason, some of you may have very limited experience in viewing stars, constellations, planets, moons, galaxies, and other celestial objects. Even for those who live under pristine skies, without the aid of binoculars and telescopes, the sky may seem large and mysterious. With modern-day technology, radio, IR, X-ray telescopes, space telescopes, and planetary spacecraft, we have greatly expanded our knowledge of the night sky. Starry Night College makes use of this technology to bring the images and information to you, the student, in an interactive, student-oriented, and self-guided exploratory nature. As you explore your night sky, you can pick and choose the objects of interest, zoom in, see what they look like, and learn more about them.

These activities should be thought of as lab experiences, not evaluation tools, where the carrying out of the activities themselves, seeing and learning what Starry Night College has to show you, is what it's all about. I hope you will enjoy your journey of exploration as much as I have.

GETTING STARTED:

If you have a Pearson textbook with a download code for Starry Night College, go to http://starrynight education. com/pearson/ to access the download page. Enter your code and type in the Captcha authorization to receive your download. If your code comes back as incorrect, make sure you have not confused the number 0 and letter o, or the number 1 and letter l. The code is also case-sensitive. If you are unable to install or run Starry Night College, or if you are experiencing any other technical issues, go to http://www.starrynightsupport.com/ support_trio.

In order for Starry Night College to run, QuickTime™ must also be installed on your computer. If QuickTime is not already installed on your computer, the Starry Night College installer will install it for you. In order to view the PDFs of the *User's Guide* and *Companion Book*, Acrobat Reader™ must also be installed. If Acrobat Reader is not already installed on your computer, the Starry Night College installer will install it for you. For Mac users, if your versions of QuickTime or Acrobat Reader are not up to date, click on the Apple menu, and choose "Software Update." For more detailed system requirements and installation information, go to www.starrynighteducation.com/college.

As you run Starry Night College for the first time, you may also be prompted to set your home location. If not, you can set or change your home location by selecting the **File** menu and then **Set Home Location**.

At this point, you should be able to proceed with the Starry Night College activities in this booklet. As you become an advanced user of the software, you can access the *User's Guide* for details and specifics on all of Starry Night College's advanced functions and capabilities. The *Companion Book* is an additional supplement that makes use of preconfigured viewing configurations accessed through the **Help menu** at the top of your screen. To access the Starry Night College *User's Guide* select **User's Guide** from the **Help menu** at the top of your screen.

You should keep in mind that you do not need to know everything about the software to get started. It's best to learn by doing. If you can load the software and set your home location, you are set to go.

MAC USERS:

The Starry Night College software will run on both PC Windows and Mac OS platforms; however, there are some important differences. A "right click" in Windows is generally equivalent to a "control click" on the Mac. Also, the Mac version of the software does not make use of the **button bar**. However, all **button bar** items can be accessed through the **menu** and **side panes** as described in the next section. Mac users will need to pay careful attention to the alternate paths shown for all **button bar** items. These activities were designed on a PC Windows platform and every effort has been made to render them compatible with both platforms. However, unanticipated differences are sure to occur.

FILE NAMES:

All **menu**, **button bar**, **control panel**, and **side pane** items are designated in bold. For example, "Click the **daylight** button on the **control panel**." In addition, sub-menu items, choices, and/or selections are also designated in bold. Sometimes menu items with sub-menu options will be separated with a forward slash. For example, "Click **file/preferences**, then select **factory defaults**." Also, when there are multiple choices, all caps are used to identify which option to select. For example, "OPTIONS **side pane**." In general, software commands are not capitalized, though they may be capitalized in the software menus themselves.

ERRATA:

Every effort has been made to present these activities to you error free. Be careful with turning on multiple databases at the same time. The same objects appearing in different databases can have slightly different coordinates, creating duplicates that can cause confusion. Be sure to check for software updates by selecting **LiveSky/Check For Program Updates** from the **menu** at the top of your screen. You should also periodically select **Live Sky/Update Comets/Asteroids/Satellites** to keep current with these object's changing orbital parameters.

NOTE TO INSTRUCTORS:

These activities are designed to meet the needs of a large and diverse student base, for both the conceptual- and math-based introductory astronomy courses. These activities offer the opportunity for instructors to customize the learning experience and/or to focus on ideas and concepts emphasized in their lecture course. Some of the tables include blank fields so that students may answer supplemental questions. For example, if an instructor is emphasizing different astronomical coordinate systems in their course, they can ask that in addition to answering the simple questions asked for in the activity, students should determine the equatorial and ecliptic coordinates as well. An instructor may also ask for students to print out information to be attached and turned in.

HOW TO ENCOURAGE STUDENTS TO DO THEIR OWN WORK:

Should copying be of concern (sometimes an issue for large introductory classes), an easy way to reduce and/or eliminate the problem is to assign a different time of the night to each student. This will force many answers to be unique (specifically the local altitude and azimuth coordinates). All objects viewed will have a unique set of coordinates that can be checked. Instead of different times of the night, an instructor may also choose to assign different days of the year, and/or assign a different home location for each student in the class. Most answers in an activity will remain common and will be easy to check for the entire class; however, should there be a concern regarding a specific student's work, the local coordinates may be checked.

REFERENCES:

Constellation stories are derived from various sources; however, most of the material presented here is derived from *Patterns in the Sky* by Julius D. W. Staal.

STARRY NIGHT COLLEGE "ON SCREEN" BUTTONS & CONTROLS:

The Starry Night College "on screen" layout is grouped into four general areas:

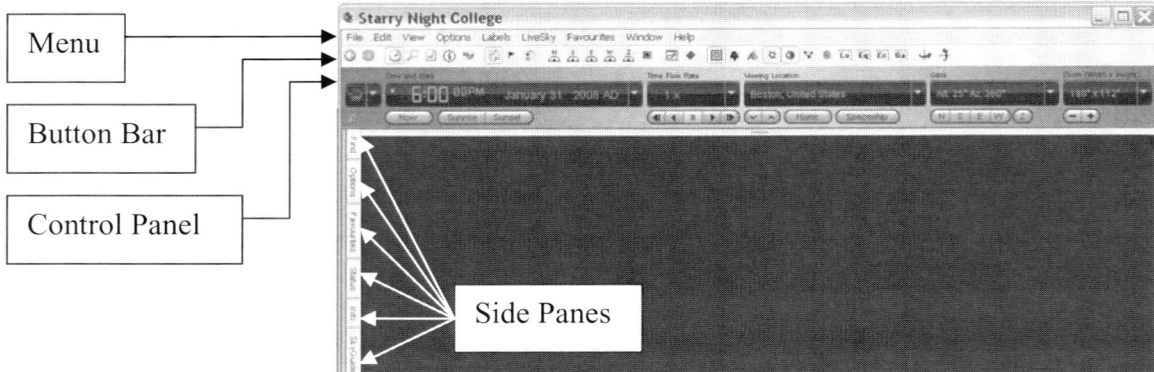

- **Menu**: Runs across the top of the screen, above the Control Panel and Button Bar. Clicking on a Menu item causes it to expand to reveal additional options.

- **Button Bar (Windows only)**: The strip of buttons located just below the Menu, found only in the Windows version of the software. Mac users can access the Button Bar items directly from the Menu.

- **Control Panel**: The strip of buttons that run horizontally below the Menu and Button Bar, and just above the main window.

- **Side Panes**: The panes or tabs that run vertically along the left side of the screen. When you click on a Side Pane, detailed settings and options become available.

The Starry Night College activities in this booklet will reference the specific buttons, controls, and functions as designated here. Please refer to your Starry Night College *User's Guide* for detailed descriptions of each function. The *User's Guide* is located on the CD and installed along with the software on your computer. You can access this document from the CD directly, or from the Starry Night College **menu** by clicking **help/user's guide**. Here we identify for you only those functions on the **control panel** and **button bar** that are most commonly used in the Starry Night College activities found in this booklet.

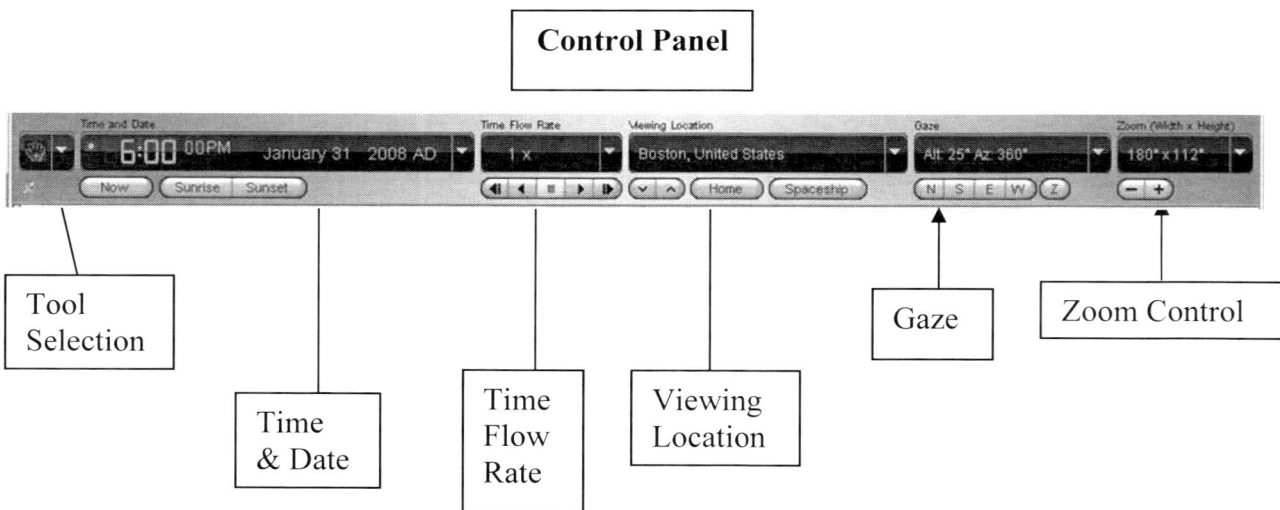

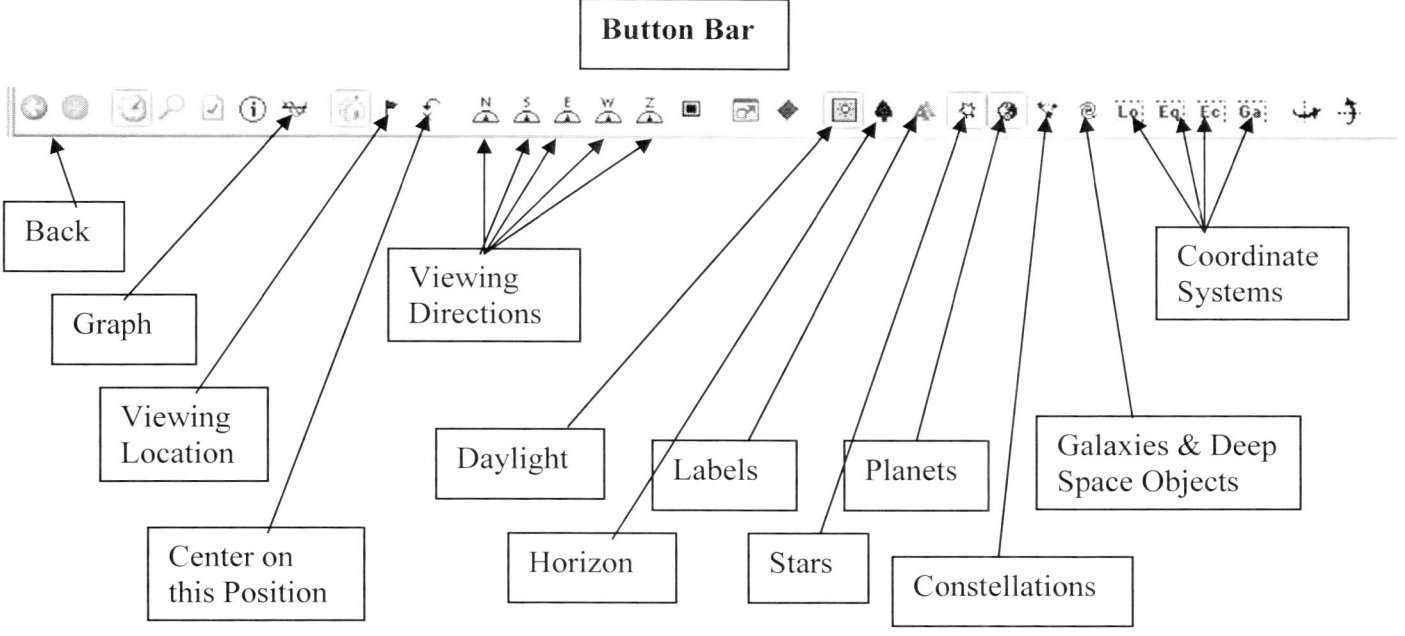

All the **button bar** items shown above can also be accessed through the **menu** as follows:
*(The Mac version of the software does not make use of the **button bar**. Mac users must use these commands.)*

- **Back:** edit/undo ...
- **Graph:** **view/show graph** or **view/hide graph**
- **Viewing Location:** **view/viewing location**
- **Center on This Position:** **edit/centre on**
- **Viewing Directions:** **click the Gaze buttons of the control panel or depress the keyboard letters N, S, E, W, and/or Z**
- **Daylight:** **view/hide daylight** or **view/show daylight**
- **Horizon:** **view/hide horizon** or **view/show horizon**
- **Labels:** **labels/show all labels** or **labels/hide all labels**
- **Stars:** **view/stars (check or uncheck the various options)**
- **Planets:** **view/planets (check or uncheck the various options)**
- **Constellations:** **view/constellations (check or uncheck the various options)**
- **Galaxies, Deep Space Objects:** **view/deep space (check or uncheck the various options)**
- **Coordinate Systems:** **view/alt/az guides, view/celestial guides, view/ecliptic guides, view/galactic guides, view/extra-galactic guides (check or uncheck the various options for each)**

The **button bar** includes various other functions, but the ones listed here are of greatest use in working with the Starry Night College activities in this booklet. You should refer to the Starry Night College *User's Guide* for detailed information on all functions.

Activity 01—EXPLORATION of the CURRENT EVENING SKY

*These activities are designed to work with the Starry Night College software that comes with your text, from any home location you choose, and with the current date and time, unless indicated otherwise. You may always revert to factory default settings by clicking **FILE/ preferences**, then selecting **factory defaults** as needed. You may also undo a command or series of commands on the PC by clicking the **back** button at the top left of the **button bar**. You should refer to the key given at the beginning of this booklet for clarification of "on screen" buttons, controls, and functions. PC **button bar** items can all be accessed through the **menu**. "Right click" on the PC is equivalent to "control click" on the Mac. All activities assume that OpenGL graphics capabilities are enabled on your computer.*

PART 1: CONSTELLATIONS

You should begin this activity at sunset. An easy way to do this is to click the drop-down menu to the right of the **date & time** field on the **control panel** at the top of your screen just above the viewing pane, and select **sunset**. Look toward the west by clicking the **W viewing direction** button located on the **button bar** across the top of your screen, or by simply keying in the letter W (Mac users should refer to the button bar commands given at the beginning of this booklet). The screen will pan toward the west.

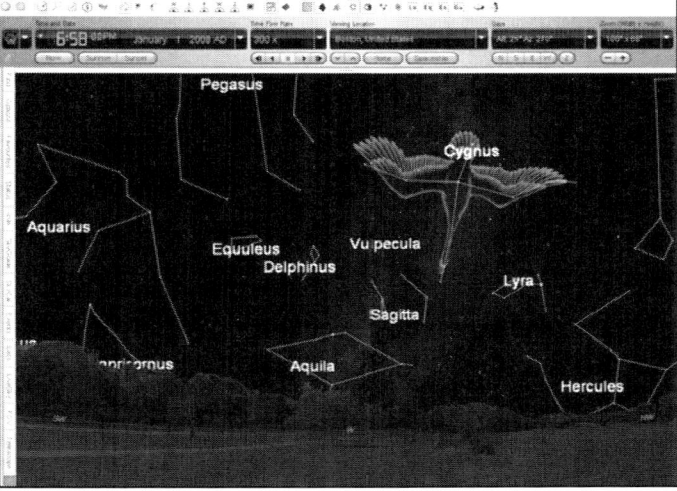

Advance time at a rate of **300×**, or 300 times faster than regular time, by clicking the drop-down menu at the right of the **time speed** field. Click the STOP **time mode** button when the sun has set, the stars have come out, and dusk is almost over.

Click on the **constellations** button found on the **button bar** at the top of the screen above the **control panel**. Right click (control click for Mac) on any constellation on the screen, careful not to be hovering directly over a specific star or object, then **select** the constellation. To look around, you can "drag" the sky by holding down the left mouse button and moving the mouse in a given direction.

Explore your current evening sky by selecting a constellation from each direction: N, S, E, W, and Z (for zenith). Identify one constellation from each direction and complete the table below. Describe the shape in your own words.

DATE:_____ TIME:_____

DIRECTION	CONSTELLATION	SHAPE
WEST		
EAST		
NORTH		
SOUTH		
ZENITH		

Given clear skies, there are some 6,000 stars visible to the unaided eye (without telescopes or binoculars). These stars are grouped into 88 constellations, often named after mythological characters and animals. On any given night you can only see some of these, depending on the time of night and your position on Earth.

PART 2: CURRENT EVENING SKY

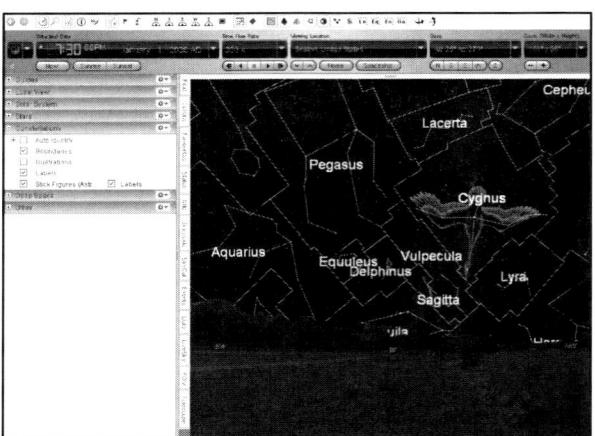

We use constellations as a way to identify *where* things are in the sky. For example, we may want to know where a planet is, or if a planet is in our current evening sky.

Although named for a characteristic pattern of stars, each constellation represents a specific area of the sky.

Click on the VIEW OPTIONS tab on the left **side pane,** and then click the plus sign (gray arrow for Mac) to expand the **constellations** section. Check the **boundaries** box to outline each constellation. Now click the FIND tab on the left **side pane** and note which planets are highlighted in bold. These planets are *up* and viewable in your current evening sky. Double-click on the name of a planet to identify and center the object on your screen. Explore your current evening sky by finding three planets (or other objects such as satellites, asteroids, or comets) and identifying what constellations they may be found in (use the table below).

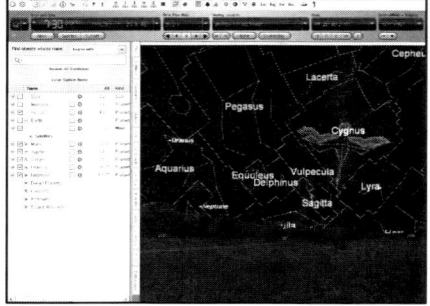

You can also describe the positions of each object by *azimuth* and *altitude*. The *azimuth* is given in degrees, as measured from North and turning clockwise. For example, N would have an *azimuth* of 0 degrees. East would have an *azimuth* of 90 degrees, South is at *azimuth* of 180 degrees, and West is at *azimuth* of 270 degrees. The *altitude* is also measured in degrees and represents the angular measure of an object above the horizon. The *azimuth* and *altitude* can be estimated by clicking on the LOCAL (lo) **coordinate system** button located on the **button bar** across the top of your screen, and referring to the degrees shown. You can also hover the pointer over the object of interest and read the precise *azimuth* and *altitude* values given. Note that the *azimuth* and *altitude* will change with time, so be careful to always record the time when taking such readings.

Find three planets (or other objects) and identify what constellations they may be found in.

DATE:_____ TIME:_____

PLANET/OBJECT	CONSTELLATION	AZIMUTH	ALTITUDE

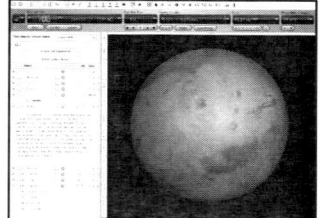

Now that you have found the planets in our current evening sky, you can take a closer look at them by zooming in with the **zoom control** buttons at the far right of the **control panel** (click the PLUS sign to zoom in), or by hovering your mouse over the planet, right clicking (control clicking for Mac), then selecting **magnify**. To return back to normal viewing, click the MINUS **zoom control** button, and zoom out as far out as you can. To learn more about a planet or object that you are viewing, click the information icon **(i)** to the right of the planet in the FIND **side pane**. When done, click on the FIND tab again to toggle back to a full screen view.

PART 3: ROTATION

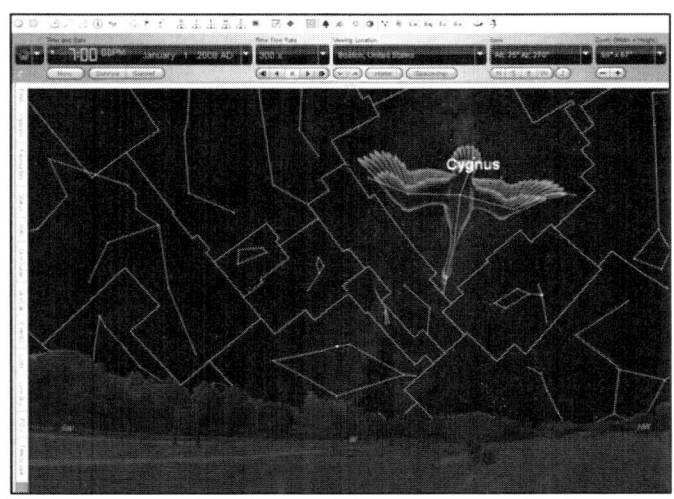

Look toward the west by clicking the **W viewing direction** button, or by simply keying in the letter W. The screen will pan toward the west. Now click on the PLAY **time mode** button to have time move forward. Select **300× time speed**. Click STOP after a few seconds of motion. Don't continue the motion for more than a few seconds. Circle the correct answer.

Do stars set: (a) straight down,
                    (b) at an angle toward the left,
                    (c) at an angle toward the right?

Now look toward the east and repeat. Click the PLAY **time mode** button for a few seconds and observe the motion. Circle the correct answer.

Do stars rise: (a) straight up,
                    (b) at an angle toward the left,
                    (c) at an angle toward the right?

Is your answer looking toward the *east* the same or different than your answer looking toward the *west*?

        same/different (circle correct answer here).

Why or why not? Write your answer in the space provided below.

We will explore this question more fully in a separate activity.

While looking toward the east, click the FIND tab on the left **side pane** once again. The highlighted objects are those viewable in your current evening sky. Now click the PLAY **time mode** button to move time forward at **300× time speed**. Let time move forward for a few hours, or until one or more new objects become highlighted. Click the STOP button.

Complete the following table for one new object and the same three objects you previously looked at in Part 2.

DATE:_____ TIME:_____

PLANET/OBJECT	CONSTELLATION	AZIMUTH	ALTITUDE

At this later time of the night, are the three planets or objects you previously looked at in the same constellations?

yes/no (circle correct answer here).

How about their *azimuths* and *altitudes*? Did they remain the same?

yes/no (circle correct answer).

Based on your discussions in class and the material in your text, can you offer an explanation why in the space below?

Activity 02—ROTATIONAL MOTION of EARTH

These activities are designed to work with the Starry Night College software that comes with your text, from any home location you choose, and with the current date and time, unless indicated otherwise. You may always revert to factory default settings by clicking FILE/ preferences, then selecting factory defaults as needed. You may also undo a command or series of commands on the PC by clicking the back button at the top left of the button bar. You should refer to the key given at the beginning of this booklet for clarification of "on screen" buttons, controls, and functions. PC button bar items can all be accessed through the menu. "Right click" on the PC is equivalent to "control click" on the Mac. All activities assume that OpenGL graphics capabilities are enabled on your computer.

PART 1: ROTATIONAL MOTION IN THE NORTHERN HEMISPHERE

(This activity assumes you are located at mid-latitudes in the northern hemisphere. If not, you should change your location accordingly. Use the viewing location button on the button bar, and select any midlatitude city in the Northern Hemisphere. Latitudes of 30–60 degrees work well.)

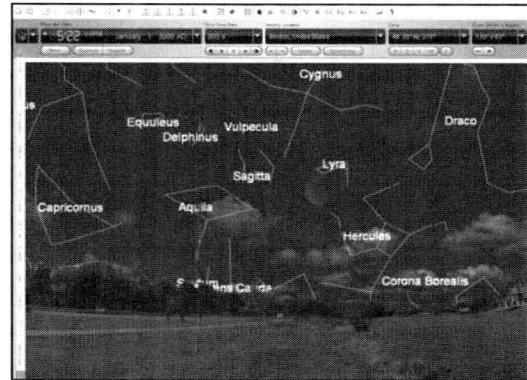

Use the drop-down menu by the **date & time** field located on the **control panel** to select **sunset**. Click on the **constellations** button located on the **button bar** to show constellations (Mac users should refer to the button bar commands given at the beginning of this booklet). Click the **W viewing direction** button on the **button bar**, or hit W on your keyboard to look toward the west. Select a **time speed** of **300×**, or 300 times faster than regular time, and keep this setting on throughout this activity. Watch the stars and constellations set. Do they set straight down, at an angle to the left, or at an angle to the right?

SETTING ANGLE (Circle the correct answer)

STRAIGHT DOWN	DOWN TOWARD LEFT	DOWN TOWARD RIGHT

If during this activity you return to daylight, simply click the **daylight** button on the **button bar** to turn daylight off. If time should stop during the activity, click the PLAY **time mode** button to start again. Now switch your field of view toward the east by clicking on the **E viewing direction** button, or selecting the E on your keyboard. Be sure time is moving forward at **300×**. Now that you are looking toward the east, in what angle do the stars and constellations move?

RISING ANGLE (Circle the correct answer)

STRAIGHT UP	UP TOWARD LEFT	UP TOWARD RIGHT

Are the rising and setting angles the same? Can you think of a reason why or why not?

Now click the **S viewing direction** button on the **button bar**, or hit the S key on your keyboard to look toward the south. Be sure time is moving forward at **300×**. What direction are the stars and constellations moving this time? Be sure to look at the motion right above the south direction and for only short distances.

RISING ANGLE (Circle the correct answer)

STRAIGHT UP	STRAIGHT DOWN	HORIZONTALLY TO THE LEFT	HORIZONTALLY TO THE RIGHT	AT A SHARP ANGLE

How about looking north? Click the **N viewing direction** button on the **button bar**, or hit the N key of your keyboard to look north. How would you describe the motion you see here? To help you visualize the motion, click the EQUATORIAL (Eq) **coordinate system** button to superimpose an equatorial coordinate system on the celestial sphere (Mac users should refer to the button bar commands given at the beginning of this booklet). The lines you see are similar to lines of latitude and longitude on the Earth, but projected out to the sky. Continue moving forward in time at **300×**, and draw a picture and show arrows indicating the motion you observe.

If the motion is circular, are the stars moving clockwise or counterclockwise?

Clockwise / Counterclockwise (circle correct answer)

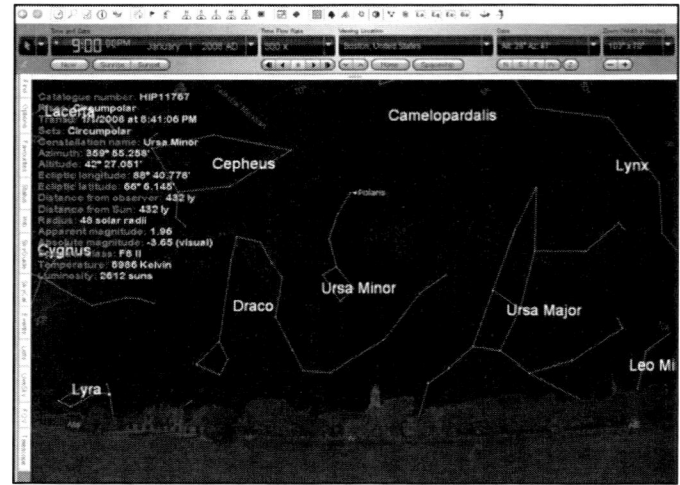

The Earth is a sphere rotating in space. The motions of stars and constellations you have been observing are caused by Earth's motion. The stars and constellations, themselves, are not moving to cause this apparent motion. The discrepancies in motion demonstrated by your different viewing angles are due to Earth's spherical shape and rotational motion. Our view toward the north clearly shows how the stars appear to move about a specific point; that point being the projection of Earth's North Pole out into space. We call this point the *North Celestial Pole (NCP)*. The constellations revolving close to this point are referred to as *circumpolar constellations*. For higher latitudes, these constellations may never set; they simply go round and round the *North Celestial Pole*. Be sure to keep the EQUITORIAL (Eq) **coordinate system** on (as described previously).

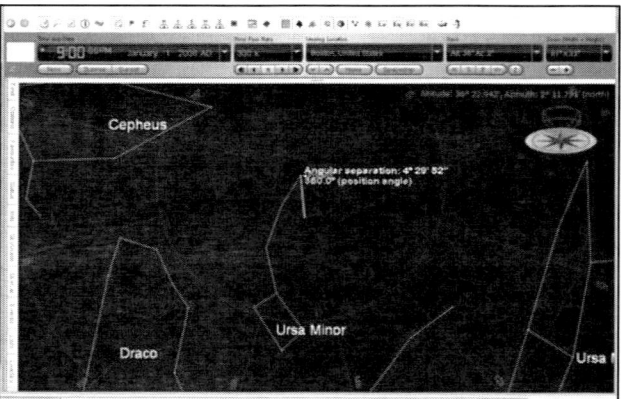

Use your mouse to point at the nearest star to the *North Celestial Pole*. Information about this star should appear on your screen. Zoom in a bit by using the **zoom control** on the upper right of the **control panel**. Determine the angular separation of this star from the *North Celestial Pole*. You can do this by hovering the mouse directly over this star (watch as the hand turns to an arrow) and then while clicking down and holding the mouse button, dragging the mouse over to the *North Celestial Pole*. Your angular separation value will be given in degrees, minutes, and seconds of arc. There are 360 degrees in a circle, 60 minutes of arc in a degree, and 60 seconds of arc in a minute. Complete the following table of information on this star.

NAME OF STAR	
TYPE (SINGLE, BINARY, or MULTIPLE)	
DISTANCE	
ANGULAR SEPARATION FROM NCP	

This star, so close to the *North Celestial Pole*, is almost directly aligned with the direction north on the Earth; thus, we call it the *North Star* (note the directions shown on the horizon). The *North Star* is not exactly aligned with the *North Celestial Pole*; however, it is within a degree of it. Also note that the *North Star* is not the brightest star in the sky. This is a common misconception. This star is not important because of its brightness, but rather because of its location. Being so close to the *North Celestial Pole*, it essentially doesn't move across the sky as other stars do, but remains fixed in the sky toward the north for all observers on Earth at all times of the year. For this reason it is a most important navigational aid, and has guided many a traveler in many a culture over the ages. To best appreciate this property of the *North Star*, select a **time speed** of **3000×**. Note that the position of the *North Star* essentially does not vary, while the rest of the sky whizzes by.

PART 2: ROTATIONAL MOTION AT THE EQUATOR

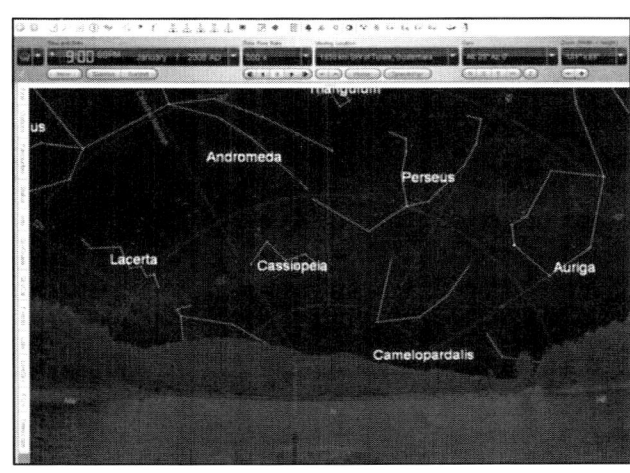

Let's investigate this further by traveling to other locations on the Earth. First, let's travel to the equator. To travel to the equator, click the **location** button on the **button bar**, select the **latitude/longitude** tab, and enter zero (0 N) for your latitude. You will then watch as you fly to your new location on Earth's equator.

Work through the same set of activities given in Part 1, and complete the following table.

LOCATION AT EQUATOR:

VIEWING DIRECTION	DIRECTION OF MOTION
WEST	
EAST	
SOUTH	
NORTH	

Explain your observations. Why do the motions appear as they do, and why so different than in Part 1?

PART 3: ROTATIONAL MOTION AT THE NORTH POLE

Let's next travel to Earth's North Pole. To travel to the North Pole, click the **viewing location** button, select the **latitude/longitude** tab, and enter 90 N for your latitude. You will then watch as you fly to the North Pole.

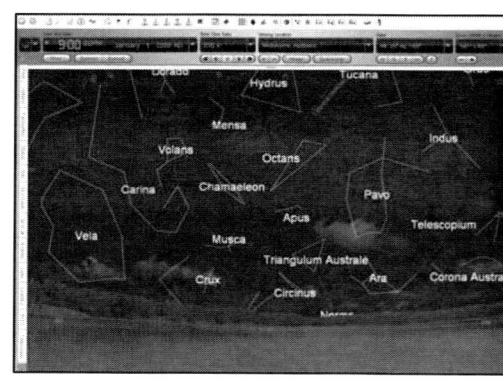

Work through the same set of activities given in Part 1, and complete the following table.

LOCATION AT NORTH POLE:

VIEWING DIRECTION	DIRECTION OF MOTION
WEST	
EAST	
SOUTH	
NORTH	

Explain your observations. Why do the motions appear as they do, and why so different than in Parts 1 and 2?

PART 4: ROTATIONAL MOTION IN THE SOUTHERN HEMISPHERE

How will Earth's rotation cause things to look different to us in the Southern Hemisphere? Use the **location** button to travel to any mid-latitude Southern Hemisphere city. Latitudes of 30–60 degrees work well.

Work through the same set of activities given in Part 1, and complete the following table.

LOCATION IN SOUTHERN HEMISPHERE:

VIEWING DIRECTION	DIRECTION OF MOTION
WEST	
EAST	
SOUTH	
NORTH	

Explain your observations. Why do the motions appear as they do? What's different compared to the motions we see from midlatitude locations in the Northern Hemisphere?

PART 5: CALCULATING RISING & SETTING ANGLES

Let's return to our home location. The simplest way to do this is to click the **location** button and select the **return home** option at the lower left.

Try and estimate the rising and setting angles you observed in Part 1. You can do this by clicking both the LOCAL **coordinate system** button as well as the EQUATORIAL **coordinate system** button, both located at the top right of the **button bar** (Mac users should refer to the button bar commands given at the beginning of this booklet). As you move time forward at **300×**, try and determine the angle of motion measured with respect to the horizontal as best you can.

Determine the latitude and rising angle for your home location (as in Part 1), the equator (as in Part 2), and the North Pole (as in Part 3), and complete the following table. The LATITUDE column values can be found by clicking the **location** button on the **button bar**, and then selecting the **latitude/longitude** tab. The RISING ANGLE column refers to the estimated rising angle as measured from the horizontal. You may need to use a protractor if you have one. Hold it to the screen as the motion is turned on and while the local and equatorial coordinate systems are showing. The setting angle (looking toward the west) should be the same, but just a mirror image. Be sure to express the angle in degrees.

When you have found the latitude and rising angle for all three locations, you then add the LATITUDE and RISING ANGLE values together to get the SUM for the last column.

LOCATION	LATITUDE	RISING ANGLE	SUM
HOME LOCATION			
EQUATOR			
NORTH POLE			

*Note that the setting angle would be the same as the rising angle (but in mirror image), and thus is not included in this table.

Do you notice a pattern? What does the SUM come out to be for all locations on Earth? Thinking of the Earth as a rotating sphere, can you outline an explanation for why this is so? You may need to draw illustrations to help explain this.

[OPTIONAL] To get more accurate values for the rising angles, you can do the following. Select the **angular separation** tool by clicking the **tool selection** icon at the far left of the **control panel**. While looking toward the west, click down and hold on any spot on the local meridian line directly over the west direction (directly over the W on the horizon). Hold and drag the cursor up along the meridian line. As you hold the button down, you will note that two numbers are displayed. The first is the angular separation. The second is the direction as measured from the *North Celestial Pole* (the red lines of the equatorial coordinate system all lead to the *North Celestial Pole*). Thus, this procedure, if done properly, can yield an accurate rising angle (the second number). Can you explain why this works? Does this work while looking in other directions? Why or why not?

Activity 03—THE PLANET MERCURY

*These activities are designed to work with the Starry Night College software that comes with your text, from any home location you choose, and with the current date and time, unless indicated otherwise. You may always revert to factory default settings by clicking **FILE/ preferences**, then selecting **factory defaults** as needed. You may also undo a command or series of commands on the PC by clicking the **back** button at the top left of the **button bar**. You should refer to the key given at the beginning of this booklet for clarification of "on screen" buttons, controls, and functions. PC **button bar** items can all be accessed through the **menu**. "Right click" on the PC is equivalent to "control click" on the Mac. All activities assume that OpenGL graphics capabilities are enabled on your computer.*

PART 1: FINDING MERCURY

You should begin this activity at Sunset. An easy way to do this is to click the drop-down menu to the right of the **date & time** field on the **control panel**, and select **Sunset**. Look toward the west by clicking the **W viewing direction** button located on the **button bar** across the top of your screen, or by simply keying in the letter W (Mac users should refer to the button bar commands given at the beginning of this booklet). The screen will pan toward the west. Select a playing speed of **300×** normal time by clicking the drop-down menu at the right of the **time speed** field. Click the STOP **time mode** button when the Sun has set, the stars have come out, and dusk is almost over. Then click on the **constellations** button to show the constellations.

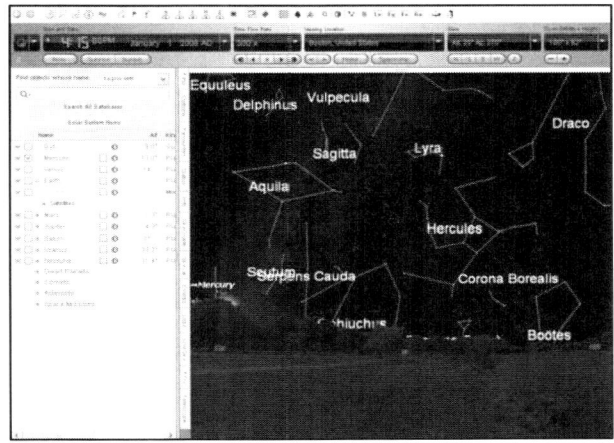

Click the FIND tab or **side pane**. A list of planets should appear. Those that are highlighted are currently up in your evening sky. Those that are not highlighted are not up in the sky at this time. We wish to find the planet Mercury. If Mercury is highlighted (see footnote if not highlighted)*, double-click or right click on it (control click for Mac) and select **center**. This will pan the screen and center Mercury. You can now zoom in on Mercury either by using the **zoom control** at the far right of the **control panel**, or by right clicking (control clicking for Mac) on the highlighted listing in the FIND **side pane**, and selecting **magnify**.

*If Mercury is not currently highlighted, you will need to move time forward to a time when Mercury will rise in the east. Start by looking toward the east by clicking the **E viewing direction** button located on the **button bar** across the top of your screen. Select a playing speed of **300** × or **3000** × normal time by clicking the drop-down menu at the right of the **time speed** field. Click the STOP **time mode** button when the planet Mercury becomes highlighted in the FIND **side pane** listing. It may be that Mercury is up during the day. If the Sun rises before Mercury does, then click the **daylight** button on the **button bar**. This will keep the sky dark so you can see the stars and constellations. Once Mercury has risen, double-click on it. This will pan the screen and center Mercury. You can now zoom in on Mercury either by using the **zoom control** at the far right of the **control panel**, or by right clicking (control clicking for Mac) on the highlighted listing in the FIND **side pane**, and selecting **magnify**.

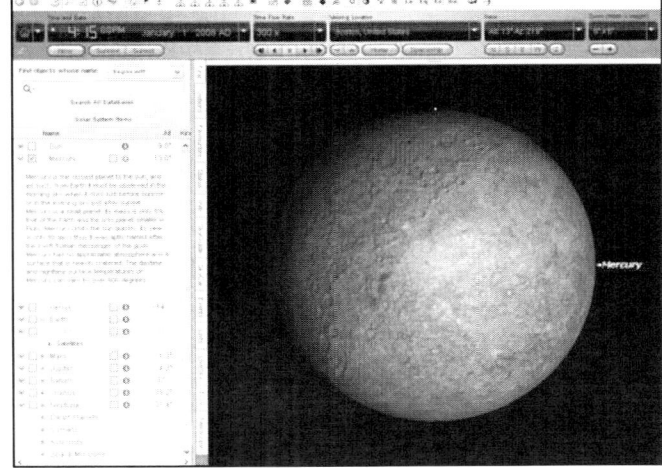

Click the information icon **(i)** in the **side pane** to read a short description of the planet Mercury. Next, select the INFO tab on the left **side pane** and click the plus sign (gray arrow for Mac) to expand the different information categories. Note that Mercury will exhibit phases depending on the position of the Sun and our viewing angle from Earth. Use the information in the OTHER DATA section to complete the following table of Mercury's physical characteristics and compare to Earth:

(You will need to know Earth's radius = 6378 km. To compute percent, divide one value by the other and multiply by 100.)

PHYSICAL CHARACTERISTICS:

Radius of Mercury:	_____ km, _____ % of Earth
Mass of Mercury:	
Angular size in arc minutes and seconds as seen from Earth:	
Orbit size (mean distance from Sun):	
Sidereal day (period of rotation):	
Solar day (noon to noon on surface):	
Length of year:	

Also complete the following table of Mercury's observational characteristics*:

OBSERVATIONAL CHARACTERISTICS:

Date, time:	
Azimuth, altitude:	
Currently in which constellation:	
Current apparent magnitude:	
Max possible magnitude as seen from Earth:	
Current distance from Earth:	
Distance at inferior conjunction (closest to Earth):	
Distance at superior conjunction (farthest from Earth):	

*For *constellation, azimuth, and altitude,* see POSITION IN SKY; for *distance from Earth,* see POSITION IN SPACE. For everything else, see OTHER DATA. To compute *distance at inferior conjunction*, take Earth's distance from Sun of 1 AU and subtract *orbit size*. To compute *distance at superior conjunction*, take Earth's distance from Sun of 1 AU and add *orbit size*.

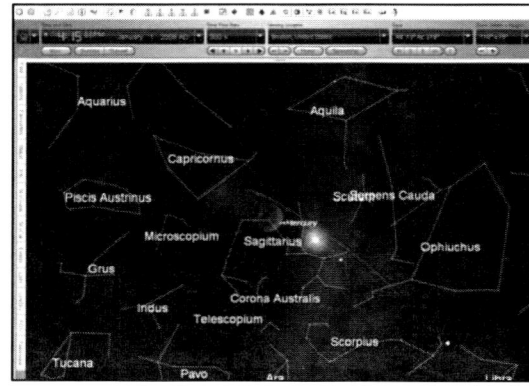

PART 2: MERCURY'S ORBITAL CHARACTERISTICS

Use the **zoom control** at the far right of the **control panel** to zoom back out to full-scale view. Right click (control click for Mac) on Mercury and select **orbit**. This shows Mercury's orbital path as seen from Earth. You will need to click the **horizon** button on the **button bar** to get rid of the Earth's horizon should it interfere with your view of Mercury's orbit. You should also click the **daylight** button to turn off daylight. You should still be locked on to Mercury. If not, right click (control click for Mac) on the planet Mercury, and then select **centre**. To maintain the proper perspective, from the **menu** select **view/ecliptic guides**, then **the ecliptic**. Next, select a **time speed** of **1 day** and click the PLAY **time mode** button. If you need to slow down or speed up, adjust the **time speed** fields as needed. You should be able to see Mercury locked in the center of your field of view, but moving across the star background. You should also be able to complete a full revolution in less than a minute. For inferior planets like Mercury and Venus, you can lock on the Sun for best results. Right click (control click for Mac) on the Sun and select **centre**.

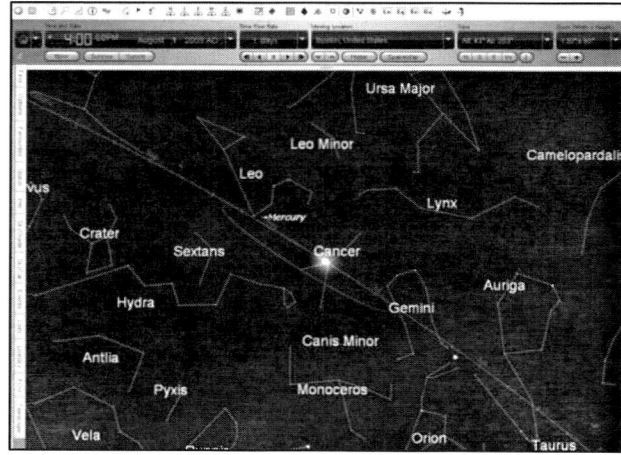

For an inferior planet such as Mercury, determine the greatest elongation and the dates of greatest elongation both for morning observing and evening observing*. The greatest elongation is the angle between the Sun and the planet, as seen from Earth. To find this value, click the pointer at the center of the Sun, then hold down the button and move the pointer to the farthest orbital extension. Take care to hover the pointer over the Sun until it changes from a hand to an arrow. If you have trouble with this, click the **tool selection** icon at the upper left of the **control panel** and select **angular separation**. Don't forget to reset **tool selection** to **adaptive** when done.

To determine the dates of greatest elongation, reset time to now and step through the orbit using the **time mode** and **time speed** settings. You may find it helpful to step forward and backward one day at a time until you have placed Mercury at its farthest extension in orbit relative to the Sun. You will need to do this for both sides of the Sun to get morning and evening greatest elongations.

Greatest elongation of Mercury (in degrees):	
Date of next occurrence:*	
Date of the following occurrence:*	

*[OPTIONAL] Can you tell if this will be viewable to us in the morning or evening? To determine if the dates of greatest elongation will occur in the morning or evening, you will need to experiment with the software. Turn the horizon back on using the **horizon** button on the **button bar** and check to see if you can view Mercury on these dates in the morning or evening.

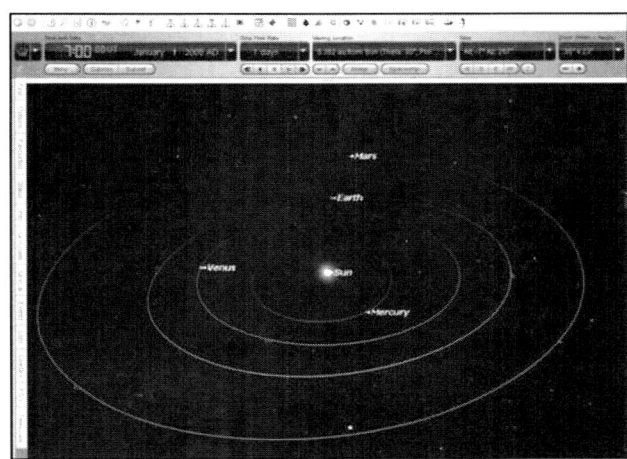

Now for the big picture: To view the position of Mercury as seen from outer space and compared to the positions of the other planets, from the **menu**, select **favourites/solar system/inner planets** and choose **inner solar system**. Indicate below which planet seems to be moving the fastest, and which planet seems to be moving the slowest around the Sun.

Fastest Planet:	
Slowest Planet:	

Click the STOP **time mode** button, select **now** from the **date and time** drop-down menu, then draw the relative positions of the planets as you see them for today's date in the space below. Then use the date determined above for greatest elongation, and draw the relative positions of the planets for that time.

Today's Date:

Greatest Elongation Date:

When done, click the **back** button repeatedly to return to Earth with your previous software configuration, or you can manually turn off planet labels and orbital tracks using the FIND **side pane**.

PART 3: FLYING TO MERCURY

Let's see what a day would be like on Mercury (technically referred to as a *solar day*). The easiest way to experience a solar day is to watch a Sunset, take note of the date and time, then watch another consecutive Sunset, and see what the time difference is between them.*

Right click (control click for Mac) on Mercury and select **go there**. You can animate the journey by first unchecking the **only animate intra planet changes** box under **file/preferences/responsiveness** (also, be sure to have your horizon turned on to see the photorealistic surface panorama). Click the **W viewing direction** button on the **button bar**, or simply hit the W key on the keyboard. Select a **time speed** of **1 day**. Click the PLAY **time mode** button, and click STOP when the Sun is near setting.** Use the FORWARD STEP and BACKWARD STEP **time mode** buttons until you see the Sun just starting to set. You may then need to select a smaller unit of time until the Sun is just touching the horizon. Note the date and time, then continue on to the next Sunset and calculate the time difference between them.

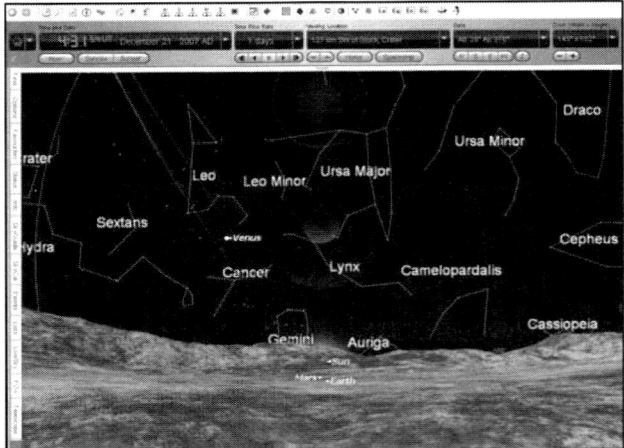

Date and time of first Sunset on Mercury:	
Date and time of second Sunset on Mercury:	
Subtract to get length of solar day:	

*For even greater accuracy, instead of using Sunset, you should use solar noon. To find solar noon, use the LOCAL (lo) **coordinate system** button to align the Sun with the local meridian line, the Sun's highest point in the sky for that day. Starry Night also gives the time of solar noon in the **date & time** drop down list on the **control panel**.

If you are unable to see the Sun to the west during Sunset, first use the **date & time drop-down menu to select **Sunset**, then click the check box to the left of the Sun in the FIND **side pane** to label the Sun, then pan to the left (SW) or right (NW) until you see it. If you still have trouble locating the Sun, right click (control click for Mac) on the Sun in the FIND **side pane** and select **center**. You will need to unlock the Sun before continuing with this activity. The easiest way to "unlock" is to simply "grab" the sky by clicking and holding the left mouse button, and then moving the mouse a little bit in any direction.

The solar day is not the same as the sidereal rotation period of a planet. This is because the solar day takes into account both the rotation of the planet and the revolution of the planet around the Sun. Your calculation of a solar day can now be compared to the sidereal day found earlier in this activity from the INFO **side pane**. Now use the table below to compare the solar day with the sidereal day, or actual rotation period of Mercury. Let's see if there is a significant difference by completing the following table.

Sidereal day of Mercury	
Solar day of Mercury	
Difference between sidereal and solar days	
Percentage of difference as compared to sidereal day*	

*To determine the percentage difference, take the *difference between sidereal and solar days*, divide it by the *sidereal day of Mercury*, and multiply by 100 to convert to a percent. This is how much longer a solar day is than a sidereal day.

How do we determine the revolutionary period, or a planet's "year"? Return to the solar system view. From the **menu**, select **favourites/solar system/inner planets** and choose **inner solar system**. Click the STOP **time mode** button, then select **now** from the **date & time** drop-down menu. Choose a **time speed** of **1 day** (or something a little slower), click the PLAY **time mode** button, and then note the time it takes for the planet to return to its original position. You may find it easiest to align the planet to the furthest left or right before starting. Complete the chart below.

Date of Mercury at start:	
Date of Mercury after one revolution:	
Subtract to get length of year:	

Activity 04—THE PLANET VENUS

*These activities are designed to work with the Starry Night software that comes with your text, from any home location you choose, and with the current date and time, unless indicated otherwise. You may always revert to factory default settings by clicking **FILE/ preferences**, then selecting **factory defaults** as needed. You may also undo a command or series of commands on the PC by clicking the **back** button at the top left of the **button bar**. You should refer to the key given at the beginning of this booklet for clarification of "on screen" buttons, controls, and functions. PC **button bar** items can all be accessed through the **menu**. "Right click" on the PC is equivalent to "control click" on the Mac. All activities assume that OpenGL graphics capabilities are enabled on your computer.*

PART 1: FINDING VENUS

You should begin this activity at Sunset. An easy way to do this is to click the drop-down menu to the right of the **date & time** field on the **control panel**, and select **Sunset**. Look toward the west by clicking the **W viewing direction** button located on the **button bar** across the top of your screen, or by simply keying in the letter W (Mac users should refer to the button bar commands given at the beginning of this booklet). The screen will pan toward the west. Select a playing speed of **300×**, or 300 times faster than regular time, by clicking the drop-down

menu at the right of the **time speed** field. Click the STOP **time mode** button when the Sun has set, the stars have come out, and dusk is almost over. Then click on the **constellations** button to show the constellations.

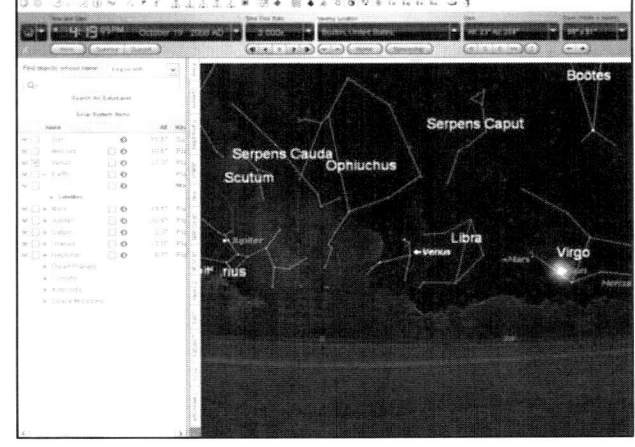

Click the FIND tab on the left **side pane**. A list of planets should appear. Those that are highlighted are currently up in your evening sky. Those that are not highlighted are not up in the sky at this time. We wish to find the planet Venus. If Venus is highlighted (see footnote if not highlighted)*, double click or right click on it (control click for Mac) and select **centre**. This will pan the screen and center Venus. You can now zoom in on Venus either by using the **zoom control** at the far right of the **control panel**, or by right clicking (control clicking for Mac) on the highlighted listing in the FIND **side pane**, and selecting **magnify**.

*If Venus is not currently highlighted, you will need to move time forward to a time when Venus will rise in the east. Start by looking toward the east by clicking the **E viewing direction** button located on the **button bar** across the top of your screen. Select a playing speed of **300×** or **3000×** normal time by clicking the drop-down menu at the right of the **time speed** field. Click the STOP **time mode** button when the planet Venus becomes highlighted in the FIND **side pane** listing. It may be that Venus is up during the day. If the Sun rises before Venus does, then click the **daylight** button on the **button bar**. This will keep the sky dark so you can see the stars and constellations. Once Venus has risen, double-click on it. This will pan the screen and center Venus. You can now zoom in on Venus either by using the **zoom control** at the far right of the **control panel**, or by right clicking (control clicking for Mac) on the highlighted listing in the FIND **side pane**, and selecting **magnify**.

Click the information icon **(i)** in the **side pane** to read a short description of the planet Venus. Next, select the INFO tab on the left **side pane** and click the plus sign (gray arrow for Mac) to expand the different information categories. Note that Venus will exhibit phases depending on the position of the Sun and our viewing angle from Earth. Use the information in the OTHER DATA section to complete the following table of Venus's physical characteristics and compare to Earth. *(You will need to know Earth's radius = 6378 km. To compute percent, divide one value by the other and multiply by 100.)*

PHYSICAL CHARACTERISTICS:

Radius of Venus:	_____ km, _____ % of Earth
Mass of Venus:	
Angular size in arc minutes and seconds as seen from Earth:	
Orbit size (mean distance from Sun):	
Sidereal day (period of rotation):	
Solar day (noon to noon on surface):	
Length of year:	

Also complete the following table of Venus's observational characteristics*:

OBSERVATIONAL CHARACTERISTICS:

Date, time:	
Azimuth, altitude:	
Currently in which constellation:	
Current apparent magnitude:	
Max possible magnitude as seen from Earth:	
Current distance from Earth:	
Distance at inferior conjunction (closest to Earth):	
Distance at superior conjunction (farthest from Earth):	

*For *constellation, azimuth, and altitude,* see POSITION IN SKY. For *distance from Earth,* see POSITION IN SPACE. For everything else, see OTHER DATA. To compute *distance at inferior conjunction,* take Earth's distance from Sun of 1 AU and subtract *orbit size.* To compute *distance at superior conjunction,* take Earth's distance from Sun of 1 AU and add *orbit size.*

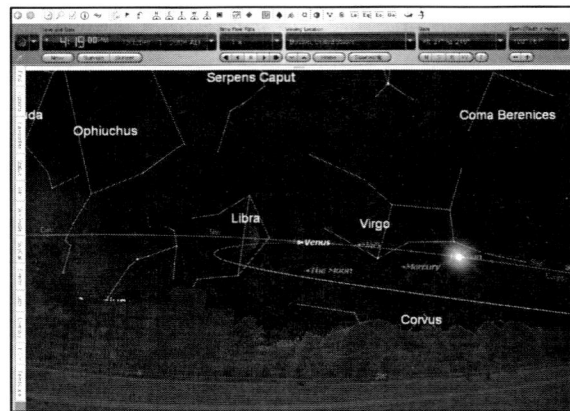

PART 2: VENUS'S ORBITAL CHARACTERISTICS

Use the **zoom control** at the far right of the **control panel** to zoom back out to full-scale view. Right click (control click for Mac) on Venus and select **orbit**. This shows Venus's orbital path as seen from Earth. You will need to click the **horizon** button on the **button bar** to get rid of the Earth's horizon should it interfere with your view of Venus's orbit. You should also click the **daylight** button to turn off daylight. You should still be locked on to Venus. If not, right click (control click for Mac) on the planet Venus, and then select **center**. To maintain the proper perspective, from the Menu select **view/ecliptic guides**, then **the ecliptic**. Next, select a **time speed** of **1 day** and click the PLAY **time mode** button. If you need to slow down or speed up, adjust the **time speed** fields as needed. You should be able to see Venus locked in the center of your field of view, but moving across the star background, and you should be able to complete a full revolution in less than a minute.

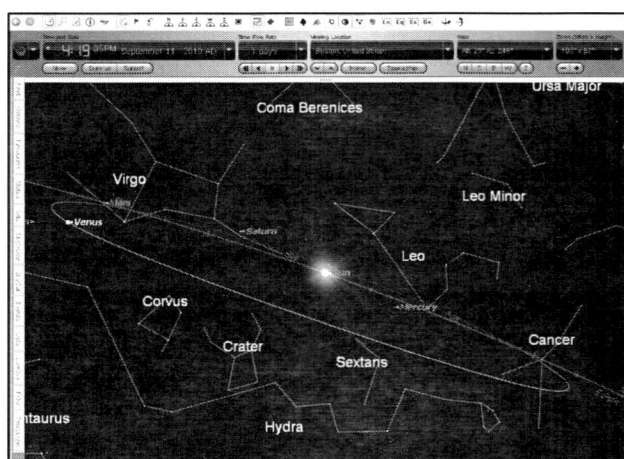

For inferior planets like Mercury and Venus, you can lock on the Sun for best results. Right click (control click for Mac) on the Sun and select **centre**.

For an inferior planet such as Venus, determine the greatest elongation and the dates of greatest elongation both for morning observing and evening observing*. The greatest elongation is the angle between the Sun and the planet, as seen from Earth. To find this value, click the pointer at the center of the Sun, then hold down the button and move the pointer to the farthest orbital extension. Be careful to hover the pointer over the Sun until it changes from a hand to an arrow. If you have trouble with this, click the **tool selection** icon at the upper left of the **control panel** and select **angular separation**. Don't forget to reset **tool selection** to **adaptive** when done.

To determine the dates of greatest elongation, reset time to now and step through the orbit using the **time mode** and **time speed** settings. You may find it helpful to step forward and backward one day at a time until you have placed Venus at its farthest extension in orbit relative to the Sun. You will need to do this for both sides of the Sun to get morning and evening greatest elongations.

Greatest elongation of Venus (in degrees):	
Date of next occurrence:*	
Date of the following occurrence:*	

*[OPTIONAL] Can you tell if this will be viewable to us in the morning or evening? To determine if the dates of greatest elongation will occur in the morning or evening, you will need to experiment with the software. Turn the horizon back on using the **horizon** button on the **button bar** and check to see if on these dates you can view Venus in the morning or evening on these dates.

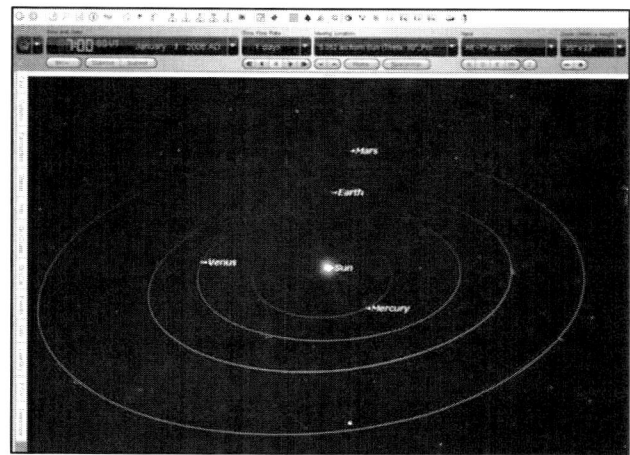

Now for the big picture: To view the position of Venus as seen from outer space and compared to the positions of the other planets, from the **menu**, select **favourites/solar system/inner planets** and choose **inner solar system**. Click the STOP **time mode** button, select **now** from the **date & time** drop-down menu, then draw the relative positions of the planets as you see them for today's date in the space below. Then use the date determined above for greatest elongation, and draw the relative positions of the planets for that time.

Today's Date:

Greatest Elongation Date:

When done, click the **back** button repeatedly to return to Earth with your previous software configuration, or you can manually turn off planet labels and orbital tracks using the FIND **side pane**.

PART 3: FLYING TO VENUS

Let's see what a day would be like on Venus (technically referred to as a *solar day*). The easiest way to experience a solar day is to watch a Sunset, take note of the date and time, then watch another consecutive Sunset, and see what the time difference is between them.*

Right click (control click for Mac) on Venus and select **go there**. You can animate the journey by first unchecking the **only animate intra planet changes** box under **file/preferences/ responsiveness** (also, be sure to have your horizon turned on to see the photorealistic surface panorama). Click the **E viewing direction** button on the **button bar**, or simply hit the E key on the keyboard (note that since Venus rotates in retrograde, Sunset will be to the east). Select a **time speed** of **1 day**. Click the PLAY **time mode** button, and click STOP when the Sun is near setting.** Use the FORWARD STEP and BACKWARD STEP **time mode** buttons until you see the Sun just starting to set. You may then need to select a smaller unit of time until the Sun is just touching the horizon. Note the date and time, then continue on to the next Sunset and calculate the time difference between them.

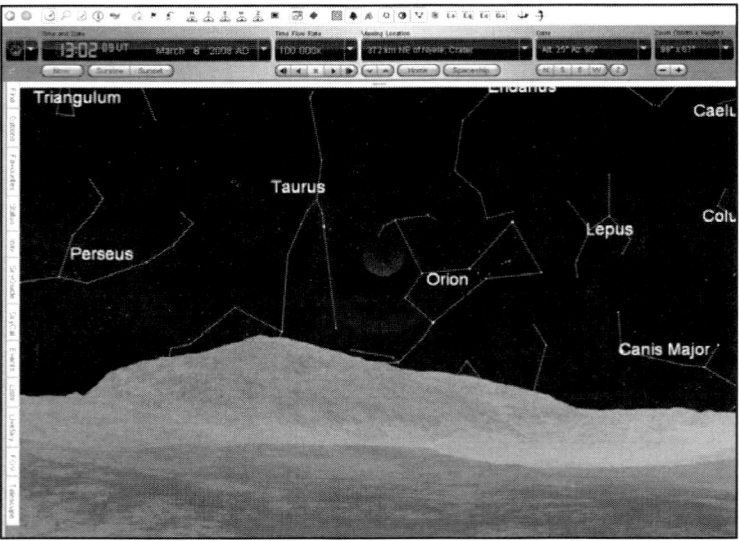

Date and time of first Sunset on Venus:	
Date and time of second Sunset on Venus:	
Subtract to get length of solar day:	

*For even greater accuracy, instead of using Sunset, you should use solar noon. To find solar noon, use the LOCAL (lo) **coordinate system** button to align the Sun with the local meridian line, the Sun's highest point in the sky for that day. Starry Night™ Pro also gives the time of solar noon in the **date & time** drop-down list on the **control panel**.

If you are unable to see the Sun to the east during Sunset, first use the **date & time drop-down menu to select **Sunset**, then click the check box to the left of the Sun in the FIND **side pane** to label the Sun, then pan to the left (NE) or right (SE) until you see it. If you still have trouble locating the Sun, right click (control click for Mac) on the Sun in the FIND **side pane** and select **centre**. You will need to unlock the Sun before continuing with this activity. The easiest way to "unlock" is to simply "grab" the sky by clicking and holding the left mouse button and then moving the mouse a little bit in any direction.

The solar day is not the same as the sidereal rotation period of a planet. This is because the solar day takes into account both the rotation of the planet and the revolution of the planet around the Sun. Your calculation of a solar day can now be compared to the sidereal day found earlier in this activity from the INFO **side pane**. Now use the table below to compare the solar day with the sidereal day, or actual rotation period of Venus. Let's see if there is a significant difference by completing the following table.

Sidereal day of Venus:	
Solar day of Venus:	
Difference between sidereal and solar days:	
Percentage of difference as compared to sidereal day*:	

*To determine the percentage difference, take the *difference between sidereal and solar days*, divide it by the *sidereal day of Venus*, and multiply by 100 to convert to a percent. This is how much longer a solar day is than a sidereal day.

How do we determine the revolutionary period, or a planet's "year"? Return to the solar system view. From the **menu**, select **favourites/solar system/inner planets** and choose **inner solar system**. Click the STOP **time mode** button, then select **now** from the **date & time** drop-down menu. Choose a **time speed** of **1 day**, click the PLAY **time mode** button, then note the time it takes for the planet to return to its original position. You may find it easiest to align the planet to the farthest left or right before starting. Complete the chart below.

Date of Venus at start:	
Date of Venus after one revolution:	
Subtract to get length of year:	

Activity 05—EARTH as SEEN from MARS and VENUS

*These activities are designed to work with the Starry Night software that comes with your text, from any home location you choose, and with the current date and time, unless indicated otherwise. You may always revert to factory default settings by clicking **FILE/preferences**, then selecting **factory defaults** as needed. You may also undo a command or series of commands on the PC by clicking the **back** button at the top left of the **button bar**. You should refer to the key given at the beginning of this booklet for clarification of "on screen" buttons, controls, and functions. PC **button bar** items can all be accessed through the **menu**. "Right click" on the PC is equivalent to "control click" on the Mac. All activities assume that OpenGL graphics capabilities are enabled on your computer.*

PART 1: VIEWING EARTH FROM MARS

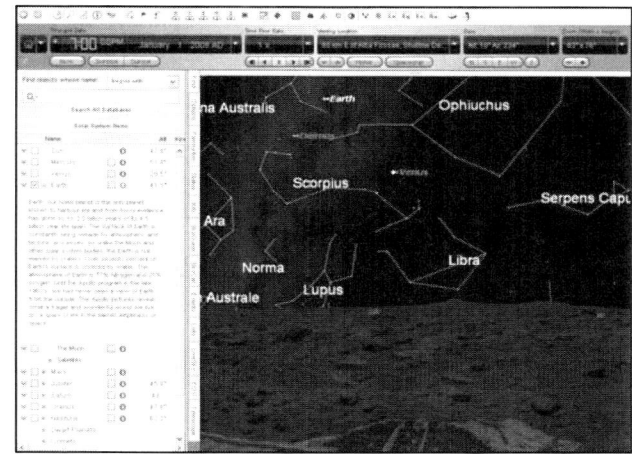

To view Earth from the planet Mars, you first need to "fly" to Mars. To better see what is happening, click the **constellations** button on the **button bar** at the top of your screen to show the constellations (Mac users should refer to the button bar commands given at the beginning of this booklet). Click the FIND tab on the left **side pane**. A list of planets should appear. Right click (control click for Mac) on **Mars**, and select **go there**. You can animate the journey by first unchecking the **only animate intra planet changes** box under **file/preferences/responsiveness** (also, be sure to have your horizon turned on to see the photorealistic surface panorama). Once you have arrived on Mars, in the FIND **side pane**, planets that are highlighted are currently up in Mars's sky as seen from your location on Mars. Those that are not highlighted are not up in Mars's sky at this time. We wish to find the planet Earth. If Earth is highlighted (see footnote if not highlighted)*, double click or right click on it (control click for Mac) and select **centre**. This will pan the screen and center Earth. You can now zoom in on Earth either by using the **zoom control** at the far right of the **control panel**, or by right clicking (control clicking for Mac) on the highlighted listing in the FIND **side pane**, and selecting **magnify**.

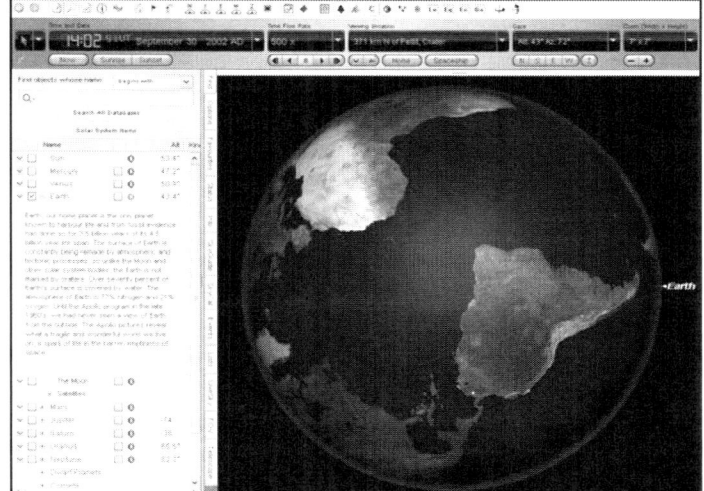

*If Earth is not currently highlighted, you will need to hide the **horizon**. If the Sun is up, then click the **daylight** button on the **button bar**. This will keep the sky dark so you can see the stars and constellations.

Click the information icon **(i)** next to the listing of Earth in the FIND tab on the left **side pane** to read a short description of Earth as seen from Mars. Next, select the INFO tab on the left **side pane** and click the plus sign (gray arrow for Mac) to expand the different information layers. Use the information in the OTHER DATA section to complete the following table of Earth's physical characteristics:

PHYSICAL CHARACTERISTICS:

Radius of Earth:	_____ km, _____ % of Earth
Mass of Earth:	
Angular size in arc minutes and seconds as seen from Mars:	
Orbit size (mean distance from Sun):	
Sidereal day (period of rotation):	
Solar day (noon to noon on surface):	
Length of year:	

Also complete the following table of Earth's observational characteristics as seen from Mars*:

OBSERVATIONAL CHARACTERISTICS:

Date, time:	
Azimuth, altitude:	
Currently in which constellation as seen from Mars:	
Current apparent magnitude as seen from Mars:	
Current distance from Mars:	
Distance at inferior conjunction (closest to Mars):	
Distance at superior conjunction (farthest from Mars):	

*For *constellation,* azimuth, *and altitude,* see POSITION IN SKY; for *distance from Mars,* see POSITION IN SPACE. For everything else, see OTHER DATA. To compute *distance at inferior conjunction,* take Mars's distance from the Sun and subtract Earth's distance from the Sun (*orbit size*). To compute *distance at superior conjunction,* take Mars's distance from the Sun and add Earth's distance from the Sun (*orbit size*).

PART 2: EARTH'S ORBITAL CHARACTERISTICS AS SEEN FROM MARS

Use the **zoom control** at the far right of the **control panel** to zoom back out to full-scale view. Right click (control click for Mac) on Earth and select **orbit**. This shows Earth's orbital path as seen from Mars. You will need to click the **horizon** button on the **button bar** to get rid of Mars's horizon should it interfere with your view of Earth's orbit. You should also click the **daylight** button to turn off daylight. You should still be locked on to Earth. If not, right click (control click for Mac) on the planet Earth, and then select **center**. To maintain the proper perspective, from the **menu**, select **view/ecliptic guides**, then **the ecliptic**. Next, select a **time speed** of **1 day** and click the PLAY **time mode** button. If you need to slow down or speed up, adjust the **time speed** fields as needed. You should be able to see Earth locked in the center of your

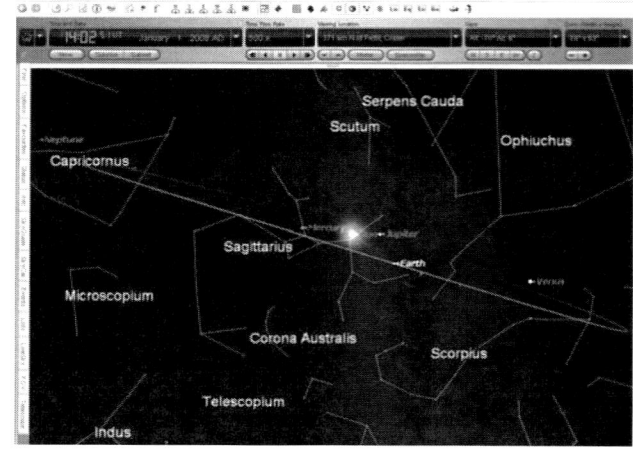

field of view, but moving across the star background. You should also be able to complete a full revolution in less than a minute. You can also try locking on the Sun. Right click (control click for Mac) on the Sun and select **centre**.

For an inferior planet such as Earth (as seen from Mars), determine the greatest elongation, and the dates of greatest elongation both for morning and evening skies*. The greatest elongation is the angle between the Sun and the planet (as seen from an observer on Mars). To find this value, click the pointer at the center of the Sun, then hold down the button and move the pointer to the farthest orbital extension. Be careful to hover the pointer over the Sun until it changes from a hand to an arrow. If you have trouble with this, click the **tool selection** icon at the upper left of the **control panel** and select **angular separation**. Don't forget to reset **tool selection** to **adaptive** when done.

To determine the dates of greatest elongation, reset time to now and step through the orbit using the **time mode** and **time speed** settings. You may find it helpful to step forward and backward one day at a time until you have placed Earth at its farthest extension in orbit relative to the Sun. You will need to do this for both sides of the Sun to get morning and evening greatest elongations.

Greatest elongation of Earth (in degrees):	
Date of next occurrence:*	
Date of the following occurrence:*	

*[OPTIONAL] Can you tell if this will be viewable to us in the morning or evening? To determine if the dates of greatest elongation will occur in the morning or evening, you will need to experiment with the software. Turn the horizon back on using the **horizon** button on the **button bar** and check to see if you view Earth in the morning or evening on these dates.

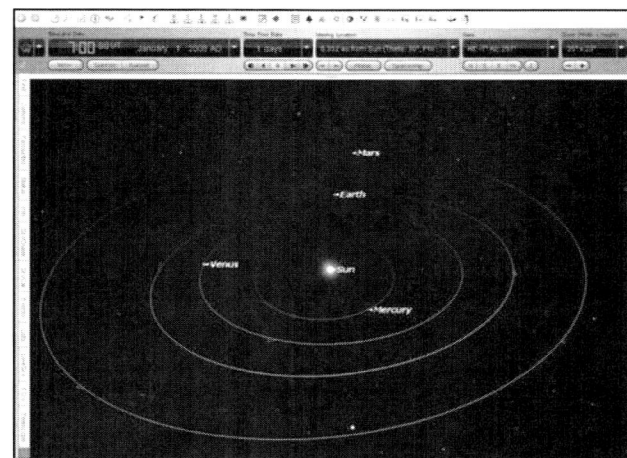

Now for the big picture: To view the position of Earth as seen from outer space and compared to the positions of the other planets, from the **menu**, select **favourites/solar system/inner planets** and choose **inner solar system**.

Click the STOP **time mode** button, select **now** from the **date and time** drop-down menu, then draw the relative positions of the planets as you see them for today's date in the space below. Then use the date determined above for greatest elongation, and draw the relative positions of the planets for that time.

Today's Date:

Greatest Elongation Date:

PART 3: EARTH AS SEEN FROM VENUS

To view Earth from the planet Venus, you will need to "fly" to Venus. First, click the **back** button repeatedly to return to Earth (or restart the program). Click the FIND tab on the left **side pane**. A list of planets should appear. Right click (control click for Mac) on **Venus** and select **go there**. (Be sure to have your horizon turned on to see the photorealistic surface panorama).

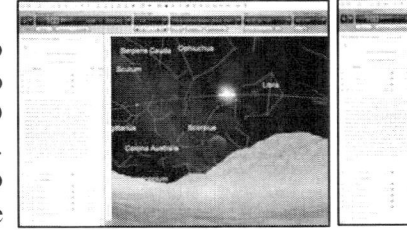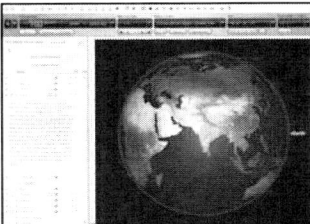

Once you have arrived on Venus, in the FIND **side pane**, planets that are highlighted are currently up in Venus's sky as seen from your location on Venus. Double-click on Earth (or you may right click and select **centre**). If Earth is not currently up in Venus's sky, you will need to hide the **horizon**. This will pan the screen and center Earth. You can now zoom in on Earth either by using the **zoom control** at the far right of the **control panel**, or by right clicking (control clicking for Mac) on the highlighted listing in the FIND **side pane**, and selecting **magnify**.

Click the INFO tab on the left **side pane** and click the plus sign (gray arrow for Mac) to expand the different information layers.

Use the information to complete the following table of Earth's observational statistics*:

Date, time:	
Azimuth, altitude:	
Currently in which constellation as seen from Venus:	
Current apparent magnitude as seen from Venus:	
Current distance from Venus:	
Distance at conjunction (aligned with the Sun):	
Distance at opposition (opposite the Sun):	

*For *constellation,* see POSITION IN SKY; for *distance from Venus,* see POSITION IN SPACE. For everything else, see OTHER DATA. To compute *distance at conjunction,* take Venus's distance from the Sun and add Earth's distance from the Sun (*orbit size*). To compute *distance at opposition,* take Earth's distance from the Sun (*orbit size*) and subtract Venus's distance from the Sun.

Use the **zoom control** at the far right of the **control panel** to zoom back out to full-scale view. Right click (control click for Mac) on Earth and select **orbit**. This shows Earth's orbital path as seen from Venus. You will need to click the **horizon** button on the **button bar** to get rid of Venus's horizon should it interfere with your view of Earth's orbit. You should also click the **daylight** button to turn off daylight. You should still be locked on to Earth. If not, right click (control click for Mac) on the planet Earth, and then select **centre**. To maintain the proper perspective, from the menu select **view/ecliptic guides**, then **the ecliptic**. Next, select a **time speed** of **1 day** and click the PLAY **time mode** button. If you need to slow down or speed up, adjust the **time speed** fields as needed. You should be able to see Earth locked in the center of your field of view, but moving across the star background. You should also be able to complete a full revolution in less than a minute.

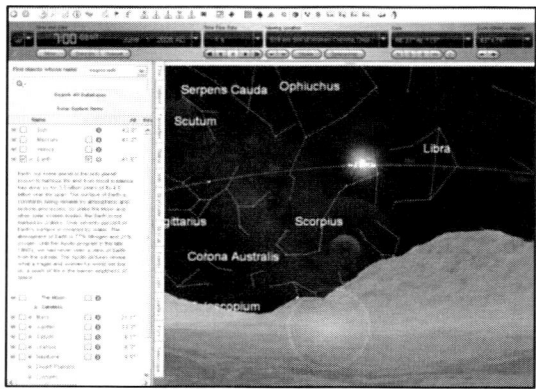

As seen from Venus, Earth behaves like a superior planet. It is of interest to determine the dates of *conjunction* and *opposition*. *Conjunction* is when the planet lines up with the Sun as viewed from an observer (aligned with the Sun). This will occur when the planet is on the far side of the Sun as compared to the observer. *Opposition* is when the planet is on the same side of the Sun as the observer, but farther out beyond the observer.

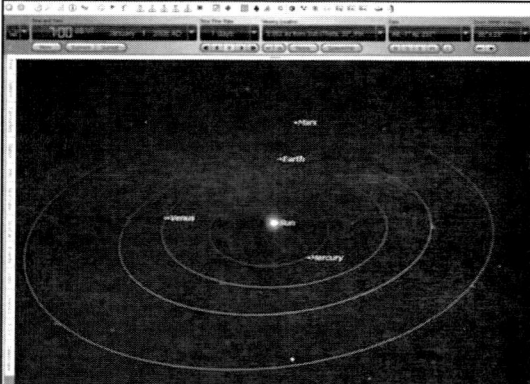

To determine the date of *conjunction,* simply step through the orbit using the **time mode** and **time speed** buttons on the **control panel** until the planet lines up with the Sun (it won't be exact, as it usually passes just above or below the Sun). You may find it helpful to step forward and backward one day at a time until you have the best alignment. Use this method to find the next date of conjunction for Earth.

Now for the solar system view: From the **menu**, select **favourites/solar system/inner planets** and choose **inner solar system.**

Let's view the conjunction event from this perspective. Advance time forward using the PLAY **time mode** button, then STOP and FORWARD STEP and BACK STEP until you are at the conjunction date you found earlier. Draw the relative positions of the terrestrial planets as you see them during this conjunction event in the space below.

DATE OF NEXT CONJUNCTION:

To determine the date of opposition, step forward one day at a time until Venus is right in between the Sun and Earth. Try and imagine a straight line from the Sun, through Venus, to Earth. This method is not very precise, but will give you a general idea of when opposition will occur.* Draw the relative positions of the terrestrial planets as you see them during this opposition event in the space below.

DATE OF NEXT OPPOSITION:

*To improve on your estimate of Earth at opposition, experiment with your **viewing location** fields to better position yourself so you can best see the opposition event. Click the **location** button on the **button bar** or, from the **menu**, experiment with **options/viewing location**. Select Sun, set your latitude for 90 degrees, and then use the **location above surface of planet** arrows on the **control panel** to launch yourself straight up until you have a face-on view of the solar system. You should turn on the orbits for Earth and Venus by clicking the FIND **side pane** and checking both boxes for each planet. First play, and then step forward and backward one day at a time to get the best possible alignment for opposition. Turning on the LOCAL (lo) **coordinate system** on the **button bar** can help with the alignment. As you move time forward, to keep from rotating with the Sun, you should select **follow Sun in orbit** from the **location control** drop-down menu on the **control panel**.

When done, click the **back** button repeatedly to return to Earth with your previous software configuration. Or you can manually turn off planet labels and orbital tracks using the FIND **side pane**, or simply restart the software.

PART 3: THE SOLAR DAY AND THE REVOLUTIONARY PERIOD (THE YEAR)

Let's determine the solar day, as we did with the other planets in other activities. Right click (control click for Mac) on Earth and select **go there** to get back to Earth (be sure to have your horizon turned on). The easiest way to experience a solar day is to watch a Sunset, take note of the date and time, then watch another consecutive Sunset, and see what the time difference is between them.*

Click the **W viewing direction** button on the **button bar**, or simply hit the W key on the keyboard. Select a **time speed** of **1 minute**. Click the PLAY **time mode** button, and click STOP when the Sun is near setting.** Use the FORWARD STEP and BACKWARD STEP **time mode** buttons until you see the Sun just starting to set. You may then need to select a smaller unit of time until the Sun is just touching the horizon. Note the date and time, then continue on to the next Sunset and calculate the time difference between them. Is this the answer you expected?

Date and time of first Sunset on Earth:	
Date and time of second Sunset on Earth:	
Subtract to get length of solar day:	

*For even greater accuracy, instead of using Sunset, you should use solar noon. To find solar noon, use the LOCAL **coordinate system** button to align the Sun with the local meridian line, the Sun's highest point in the sky for that day. Starry Night™ Pro also gives the time of solar noon in the **date & time** drop-down menu on the **control panel**.

If you are unable to see the Sun to the west during Sunset, first use the **date & time drop-down menu to select **Sunset**, then click the check box to the left of the Sun in the FIND **side pane** to label the Sun, then pan to the left (SW) or right (NW) until you see it. If you still have trouble locating the Sun, right click (control click for Mac) on the Sun in the FIND **side pane** and select **centre**. You will need to unlock the Sun before continuing with this activity. The easiest way to "unlock" is to simply "grab" the sky by clicking and holding the left mouse button and then moving the mouse a little bit in any direction.

The solar day is not the same as the sidereal rotation period of a planet. This is because the solar day takes into account both the rotation of the planet and the revolution of the planet around the Sun. Now use the table below to compare the solar day with sidereal rotation, or actual rotation of Earth. Use the value for sidereal rotation found earlier in this activity from the INFO **side pane**. The difference is small (only a few minutes), so be careful with your numbers.

Sidereal day of Earth:	
Solar day of Earth:	
Difference between sidereal and solar days:	
Percentage of difference as compared to sidereal day*:	

*To determine the percentage difference, take the *difference between sidereal and solar days*, divide it by the *sidereal day of Earth*, and multiply by 100 to convert to a percent. This is how much longer a solar day is than a sidereal day.

How do we determine the revolutionary period, or the planet's "year"?

Return to the solar system view. From the **menu**, select **favourites/solar system/inner planets** and choose **inner solar system**. Click the STOP button on the **time mode**, then select **now** from the **date and time** drop-down menu. Choose a **time speed** of **1 day**, click the PLAY **time mode** button, and then note the time it takes for the planet to return to its original position. You may find it easiest to align the planet to the farthest left or right before starting. Complete the chart below.

Date of Earth at start:	
Date of Earth after one revolution:	
Subtract to get length of year:	

Activity 06—THE PLANET MARS

These activities are designed to work with the Starry Night software that comes with your text, from any home location you choose, and with the current date and time, unless indicated otherwise. You may always revert to factory default settings by clicking FILE / preferences, then selecting factory defaults as needed. You may also undo a command or series of commands on the PC by clicking the back button at the top left of the button bar. You should refer to the key given at the beginning of this booklet for clarification of "on screen" buttons, controls, and functions. PC button bar items can all be accessed through the menu. "Right click" on the PC is equivalent to "control click" on the Mac. All activities assume that OpenGL graphics capabilities are enabled on your computer.

PART 1: FINDING MARS

You should begin this activity at sunset. An easy way to do this is to click the drop-down menu to the right of the **date & time** field on the **control panel**, and select **sunset**. Look toward the west by clicking the **W viewing direction** button located on the **button bar** across the top of your screen, or by simply keying in the letter W (Mac users should refer to the button bar commands given at the beginning of this booklet). The screen will pan toward the west. Select a playing speed of **300×** normal time by clicking the drop-down menu at the right of the **time speed** field. Click the STOP **time mode** button when the Sun has set, the stars have come out, and dusk is almost over. Then click on the **constellations** button to show the constellations.

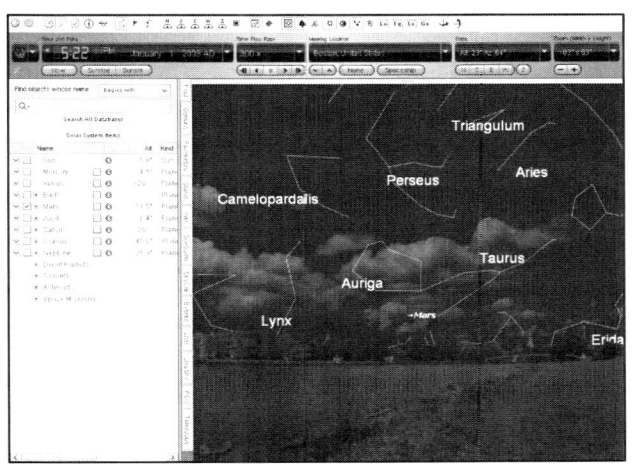

Click the FIND tab on the left **side pane**. A list of planets should appear. Those that are highlighted are currently up in your evening sky. Those that are not highlighted are not up in the sky at this time. We wish to find the planet Mars. If Mars is highlighted (see footnote if not highlighted)*, double-click or right click on it (control click for Mac) and select **centre**. This will pan the screen and center Mars. You can now zoom in on Mars either by using the **zoom control** at the far right of the **control panel**, or by right clicking (control clicking for Mac) on the highlighted listing in the FIND **side pane** and selecting **magnify**.

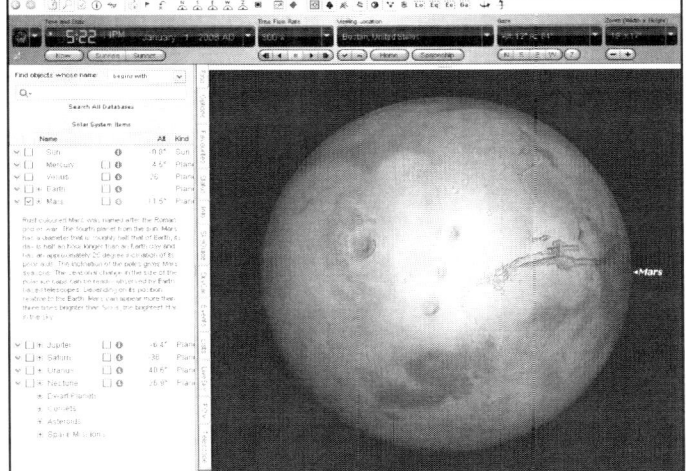

*If Mars is not currently highlighted, you will need to move time forward to a time when Mars will rise in the east. Start by looking toward the east by clicking the **E viewing direction** button located on the **button bar** across the top of your screen. Select a playing speed of **300×** or **3000×** normal time by clicking the drop-down menu at the right of the **time speed** field. Click the STOP **time mode** button when the planet Mars becomes highlighted in the FIND **side pane** listing. It may be that Mars is up during the day. If the Sun rises before Mars does, then click the **daylight** button on the **button bar**. This will keep the sky dark so you can see the stars and constellations. Once Mars has risen, double-click on it. This will pan the screen and center Mars. Zoom all the way in on Mars either by using the **zoom control** at the far right of the **control panel**, or by right clicking (control clicking for Mac) on the highlighted listing in the FIND **side pane** and selecting **magnify**.

Click the information icon **(i)** in the **side pane** to read a short description of the planet Mars. Next, select the INFO tab on the left **side pane** and click the plus sign (gray arrow for Mac) to expand the different information categories. Use the information in the OTHER DATA section to complete the following table of Mars's physical characteristics and compare to Earth.

(You will need to know Earth's radius = 6378 km. To compute percent, divide one value by the other and multiply by 100.)

PHYSICAL CHARACTERISTICS:

Radius of Mars:	_____ km, _____ % of Earth
Mass of Mars:	
Angular size in arc minutes and seconds as seen from Earth:	
Orbit size (mean distance from the Sun):	
Sidereal day (period of rotation):	
Solar day (noon to noon on surface):	
Length of year:	

Also complete the following table of Mars' observational characteristics*:

OBSERVATIONAL CHARACTERISTICS:

Date, time:	
Azimuth, altitude:	
Currently in which constellation:	
Current apparent magnitude:	
Max possible magnitude as seen from Earth:	
Current distance from Earth:	
Distance at conjunction (aligned with the Sun):	
Distance at opposition (opposite the Sun):	

*For *constellation, azimuth, and altitude,* see POSITION IN SKY; for *distance from Earth,* see POSITION IN SPACE. For everything else, see OTHER DATA. To compute *distance at conjunction,* take *orbit size* and add Earth's distance from the Sun of 1 AU. To compute *distance at opposition,* take *orbit size* and subtract Earth's distance from Sun of 1 AU.

Hovering the mouse over the Martian surface will provide geographical information. Rotate Mars by selecting **3000× time speed,** and click the **daylight** and **horizon** buttons off on the control panel. Using the maps and pictures shown in your textbook, as Mars rotates, see if you can find Valles Marineris (4000 km long, 190 km wide, and 6 km deep). Also, look for Olympus Mons, the tallest volcano in the solar system, rising 26 km (16 mi) above the Martian surface. Look for either of the polar ice caps.* What other features can you find? Locate four other features discussed in your text and identify them in the table below, noting their approximate location with respect to the center of the planet (for the specific date and time Starry Night™ Pro indicates).**

GEOGRAPHICAL FEATURES:

FEATURE	DATE	TIME	WHERE on MARS (N, S, E, W)
Valles Marineris			
Olympus Mons			
North Polar Ice Cap*			
South Polar Ice Cap*			

**Depending on the time of year, Mars may be tilted slightly toward us or away from us, making one or possibly both of the polar ice caps difficult to see.

*Be careful to get properly oriented. Depending on your viewing angle from Earth, Mars's axis of rotation may be tilted. Look for the polar ice cap (if visible) to help get you oriented, and cross reference with diagrams in your text. You may also click the **viewing location** button on the **button bar**, select Mars from the drop-down menu at the top right, then click on the **map** tab for a properly oriented view of Mars.

PART 2: MARS'S ORBITAL CHARACTERISTICS

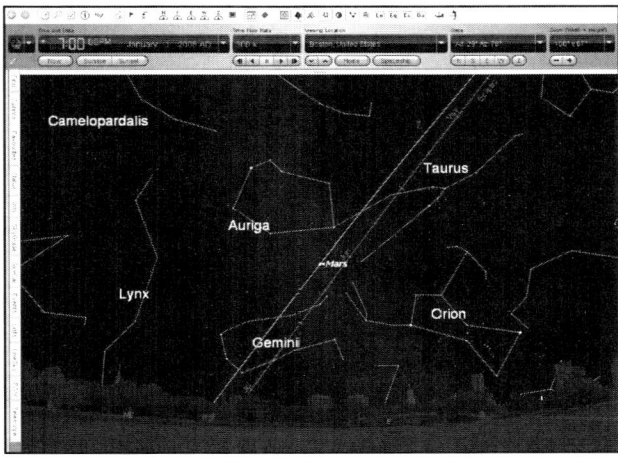

Use the **zoom control** at the far right of the **control panel** to zoom back out to full-scale view. Right click (control click for Mac) on Mars and select **orbit**. This shows Mars's orbital path as seen from Earth. If you haven't done so already, you will need to click the **horizon** button on the **button bar** to get rid of the Earth's horizon should it interfere with your view of Mars's orbit. You should also click the **daylight** button to turn off daylight. You should still be locked on to Mars. If not, right click (control click for Mac) on the planet Mars, then select **center**. To maintain the proper perspective, from the menu, select **view/ecliptic guides**, then **the ecliptic**. Next, select a **time speed** of **1 day** and click the PLAY **time mode** button. If you need to slow down or speed up, adjust the **time speed** fields as needed. You should be able to see Mars locked in the center of your field of view, but moving across the star background. You should also be able to complete a full revolution in less than a minute.

For a superior planet such as Mars, it is of interest to determine the dates of *conjunction* and *opposition*. *Conjunction* is when the planet lines up with the Sun as viewed from Earth. This will occur when the planet is on the opposite side of the Sun as compared to us on Earth. *Opposition* is when the planet is on the same side of the Sun as us, but beyond the Earth so that it appears high in our night sky around midnight. This is the best position for observing the planet (being high in the night sky and positioned closest to the Earth due to its orbit).

To determine the date of *conjunction*, first return to the present time by inputting today's date in the **date & time** field, then simply step through the orbit using the **time mode** and **time speed** buttons on the **control panel** until the planet lines up with the Sun (it won't be exact, as it usually passes just above or below the Sun). You may find it helpful to step forward and backward one day at a time until you have the best alignment. Use this method to find the next date of conjunction for Mars.

Date of Mars at next conjunction:	

Now for the big picture: To view the position of Mars as seen from outer space and compared to the positions of the other planets, from the **menu**, select **favourites/solar system/inner planets** and choose **inner solar system**. Indicate below which planet seems to be moving the fastest, and which planet seems to be moving the slowest around the Sun.

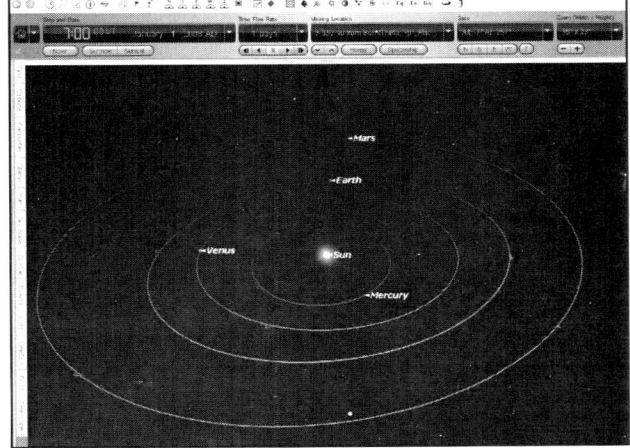

Fastest planet:	
Slowest planet:	

Click the STOP **time mode** button, select **now** from the **date and time** drop-down menu, then draw the relative positions of the planets as you see them for today's date in the space on the next page.

MARS TODAY:

Let's view the conjunction event from this solar system perspective. Advance time forward using the PLAY **time mode** button, then STOP and FORWARD STEP and BACK STEP until you are at the conjunction date you found earlier. Draw the relative positions of the Earth and Mars as you see them during this conjunction event in the space below.

MARS AT CONJUNCTION:

To determine the date of opposition, first return to the present time by inputting today's date in the **date & time** field, step forward one day at a time until the Earth is right in between the Sun and Mars. Try and imagine a straight line from the Sun, through Earth, to Mars. This method is not very precise, but will give you a general idea of when opposition will occur.*

Date of Mars at opposition:	

Draw the relative positions of the planets as you see them during this opposition event in the space below.

MARS AT OPPOSITION:

*To improve on your estimate of Mars at opposition, experiment with your **viewing location** fields to better position yourself so you can best see the opposition event. Click the **location** button on the **button bar**, or from the **menu** experiment with **options/viewing location**. Select Sun, set your latitude for 90 degrees, and then use the **location above surface of planet** arrows on the **control panel** to launch yourself straight up until you have a face-on view of the solar system. You should turn on the orbits for Earth and Mars by clicking the FIND **side pane** and checking both boxes for each planet. First play, and then step forward and backward one day at a time to get the best possible alignment for opposition. Turning on the LOCAL (lo) **coordinate system** on the **button bar** can help with the alignment. As you move time forward, to keep from rotating with the Sun, you should select **follow Sun in orbit** from the **location control** drop-down menu on the **control panel**.

When done, click the **back** button repeatedly to return to Earth with your previous software configuration, or you can manually turn off planet labels and orbital tracks using the FIND **side pane**.

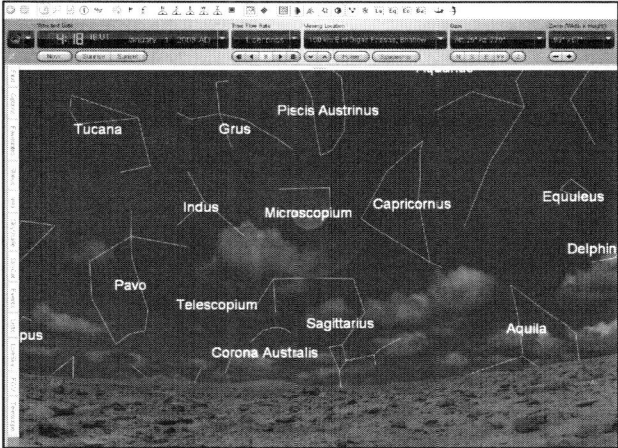

PART 3: FLYING TO MARS

Let's see what a day would be like on Mars (technically referred to as a *solar day*). The easiest way to experience a solar day is to watch a sunset, take note of the date and time, then watch another consecutive sunset, and see what the time difference is between them.*

Right click (control click for Mac) on Mars and select **go there**. You can animate the journey by first un-checking the **only animate intra planet changes** box under **file/preferences/responsiveness** (also, be sure to have your horizon turned on to see the photorealistic surface panorama). Click the **W viewing direction** button on the **button bar**, or simply hit the W key on the keyboard. Select a **time speed** of **1 minute**. Click the PLAY **time mode** button, and click STOP when the Sun is near setting.** Use the FORWARD STEP and BACKWARD STEP **time mode** buttons until you see the Sun just starting to set. You may then need to select a smaller unit of time until the Sun is just touching the horizon. Note the date and time, then continue on to the next sunset and calculate the time difference between them.

Date and time of first sunset on Mars:	
Date and time of second sunset on Mars:	
Subtract to get length of solar day:	

*For even greater accuracy, instead of using sunset, you should use solar noon. To find solar noon, use the LOCAL (lo) **coordinate system** button to align the Sun with the local meridian line, the Sun's highest point in the sky for that day. Starry Night™ Pro also gives the time of solar noon in the **date & time** drop-down menu on the **control panel**.

If you are unable to see the Sun to the west during sunset, first use the **date & time drop-down menu to select **sunset**, then click the check box to the left of the Sun in the FIND **side pane** to label the Sun, then pan to the left (SW) or right (NW) until you see it. If you still have trouble locating the Sun, right click (control click for Mac) on the Sun in the FIND **side pane** and select **centre**. You will need to unlock the Sun before continuing with this activity. The easiest way to "unlock" is to simply "grab" the sky by clicking and holding the left mouse button and then moving the mouse a little bit in any direction.

The solar day is not the same as the sidereal rotation period of a planet. This is because the solar day takes into account both the rotation of the planet and the revolution of the planet around the Sun. Your calculation of a solar day is, therefore, a little different than the sidereal day found earlier in this activity from the INFO **side pane**.

How do we determine the revolutionary period, or a planet's "year"?

Return to the solar system view. From the **menu**, select **favourites/solar system/inner planets** and choose **inner solar system**. Click the STOP **time mode** button, then select **now** from the **date & time** drop-down menu. Choose a **time speed** of **1 day**, click the PLAY **time mode** button, then note the time it takes for the planet to return to its original position. You may find it easiest to align the planet to the farthest left or right before starting. Complete the chart below.

Date of Mars at start:	
Date of Mars after one revolution:	
Subtract to get length of year:	

Activity 07—RETROGRADE MOTION of MARS

*These activities are designed to work with the Starry Night software that comes with your text, from any home location you choose, and with the current date and time, unless indicated otherwise. You may always revert to factory default settings by clicking **FILE/ preferences**, then selecting **factory defaults** as needed. You may also undo a command or series of commands on the PC by clicking the **back** button at the top left of the **button bar**. You should refer to the key given at the beginning of this booklet for clarification of "on screen" buttons, controls, and functions. PC **button bar** items can all be accessed through the **menu**. "Right click" on the PC is equivalent to "control click" on the Mac. All activities assume that OpenGL graphics capabilities are enabled on your computer.*

PART 1: MARS IN YOUR CURRENT EVENING SKY

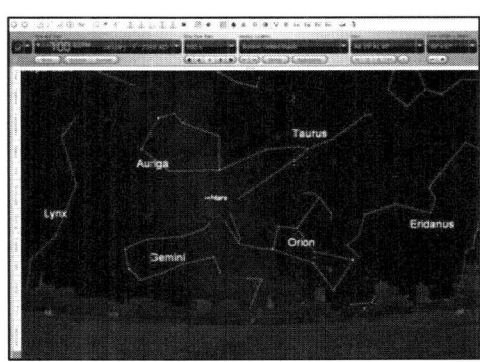

Use the drop-down menu by the **date & time** field on the **control panel** to select **sunset**. Click on the **constellations** button on the **button bar** to outline the constellations (Mac users should refer to the button bar commands given at the beginning of this booklet).

Click the FIND tab on the left **side pane** to see a list of planets. If Mars is highlighted, than it is in the current evening sky; if not, then Mars is still below the horizon. Double-click on Mars to center it in your field of view. If Mars is below the horizon, you will be given a choice to either select **best time**, or to **hide horizon**. Select **hide horizon** to make the horizon transparent, and center Mars in your field of view. Click on the information icon **(i)** in the FIND **side pane** to get a detailed description of Mars. Use the **zoom control** at the far right of the **control panel** to zoom in, or right click (control click for Mac) and select **magnify**. By clicking the INFO tab on the left **side pane**, you can get even more information on this planet.

Select **view/ecliptic guides**, then **the ecliptic**, to lock your viewing angle relative to the Earth's rotation, then select the **300× time speed** to watch Mars rotate as it moves through space. If you watch it long enough, Sunrise will lighten the sky and interfere with your viewing. Turn the Sun off by clicking the **daylight** button.

PART 2: RETROGRADE MOTION of MARS

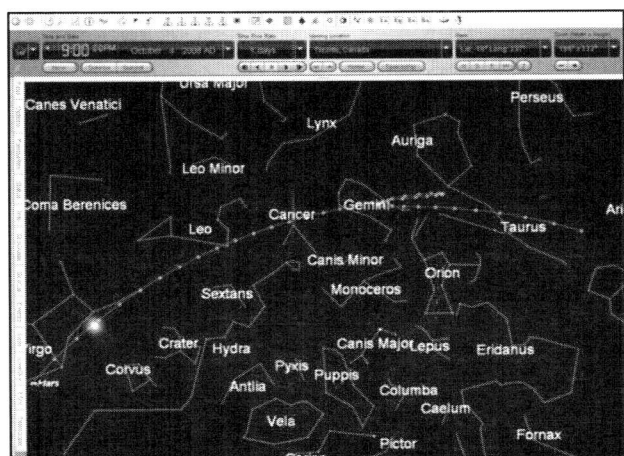

Using the **zoom control** at the upper right of the **control panel**, zoom all the way out. Animate time forward by one day at a time. Do this by using the **time speed** drop-down arrow to select **1 day**, and then click the PLAY **time mode** button. Unless already in the midst of a retrograde cycle, you should see Mars moving across the sky from right to left (prograde), yet remain centered in your screen. Watch Mars for some time and you should notice that Mars will eventually slow down relative to the background stars and constellations, then reverse for a short period of time (referred to as retrograde motion), then move forward (or prograde) again.

To help you visualize retrograde motion, press STOP **time mode** button, hover your mouse over Mars, right click (control click for Mac), and select **celestial path**. This option will draw out the path Mars takes across the sky (or celestial sphere), showing clearly how Mars reverses direction for a short time during retrograde motion. Bring time back to the present using the **date & time** settings on the **control panel** and record the details of the next three retrograde motion events. Note the dates shown along the retrograde paths to help you complete the following table. To get the exact dates, step through the retrograde motion cycle using the LEFT STEP and RIGHT STEP **time mode** buttons on the **control panel** (next to the FORWARD PLAY and BACKWARD PLAY buttons), then use the date shown. Mars' retrograde motion can also be demonstrated via **favourites/solar system/inner planets/mars retrograde**.

EVENT	Date of Onset of Retrograde Motion	Date of Onset of Prograde Motion
1		
2		
3		

Outline the three paths you observed, being careful to give dates and maintain the correct viewing perspective (angle of the orbital path as shown on your screen) in each case.

RETROGRADE EVENT 1:

RETROGRADE EVENT 2:

RETROGRADE EVENT 3:

PART 3: SOLAR SYSTEM VIEW OF RETROGRADE MOTION

Retrograde motion can be described as an "illusion," as viewed from Earth, caused by one planet passing another planet. To view retrograde motion from a solar system perspective, select **favourites/solar system/inner planets** and choose **inner solar system**. This shows a view of our solar system as seen from outer space. Observe the planets as they move around the Sun. Click the STOP **time mode** button, and then enter today's date. Don't use the **now** drop-down option since that will reset the time step. Your time step should be at **1 day**.

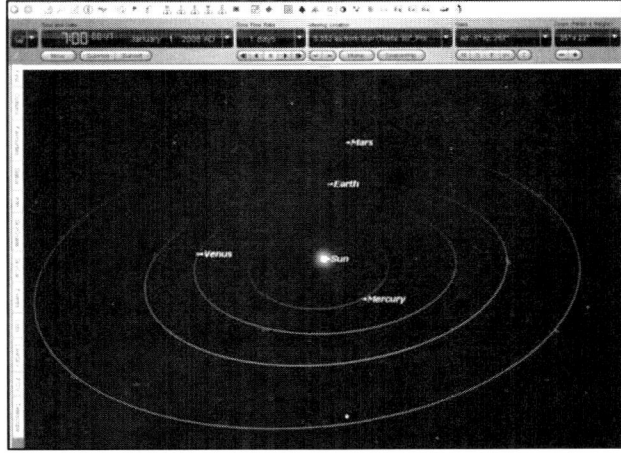

Note the relative positions of the planets. Allow time to move forward and stop around the time of the first onset of retrograde motion as given in your table in Part 2. Step through the retrograde cycle and observe what happens from this perspective and compare to what you saw with respect to the position of the Earth. Do this with the second and third retrograde events you identified. In your own words, describe and explain what is happening as you see it animated on your screen. How does this explain the "illusion" of retrograde motion as seen from Earth?

PART 4: RETROGRADE MOTION OF OTHER PLANETS AND PLUTO

As a follow up to the above exercises, use the same procedure to see if the Sun, Moon, other planets, and Pluto in our solar system exhibit retrograde motion. Complete the following table and optional diagrams on the next page.*

OBJECT	YES/NO	OBJECT	YES/NO
Sun		Jupiter	
Moon		Saturn	
Mercury		Uranus	
Venus		Neptune	
Mars	*yes*	Pluto	

*Be sure to do both this table and the optional diagrams on the next page at the same time if instructed to do both.

[OPTIONAL DIAGRAMS]

Outline the paths you observe for the planets (and Pluto) that exhibit retrograde motion, being careful to give dates and maintain the correct viewing perspective as compared to Earth's rotation by having the **view/ecliptic guides**, then **the ecliptic** option selected. Write N/A (not applicable) if retrograde motion is not observed.

Object: _____ **SUN** _____

Start of Retrograde Motion:

_____

End of Retrograde Motion:

_____

Object: _____ **JUPITER** _____

Start of Retrograde Motion:

_____

End of Retrograde Motion:

_____

Object: _____ **MOON** _____

Start of Retrograde Motion:

_____

End of Retrograde Motion:

_____

Object: _____ **SATURN** _____

Start of Retrograde Motion:

_____

End of Retrograde Motion:

_____

Object: _____ **MERCURY** _____

Start of Retrograde Motion:

_____

End of Retrograde Motion:

_____

Object: _____ **URANUS** _____

Start of Retrograde Motion:

_____

End of Retrograde Motion:

_____

Object: _____ **VENUS** _____

Start of Retrograde Motion:

_____

End of Retrograde Motion:

_____

Object: _____ **NEPTUNE** _____

Start of Retrograde Motion:

_____

End of Retrograde Motion:

_____

Object: _____ **MARS** _____

Start of Retrograde Motion:

_____

End of Retrograde Motion:

_____

Object: _____ **PLUTO** _____

Start of Retrograde Motion:

_____

End of Retrograde Motion:

_____

Activity 08—THE MARTIAN MOONS PHOBOS & DEIMOS

*These activities are designed to work with the Starry Night software that comes with your text, from any home location you choose, and with the current date and time, unless indicated otherwise. You may always revert to factory default settings by clicking **FILE/preferences**, then selecting **factory defaults** as needed. You may also undo a command or series of commands on the PC by clicking the **back** button at the top left of the **button bar**. You should refer to the key given at the beginning of this booklet for clarification of "on screen" buttons, controls, and functions. PC **button bar** items can all be accessed through the **menu**. "Right click" on the PC is equivalent to "control click" on the Mac. All activities assume that OpenGL graphics capabilities are enabled on your computer.*

PART 1: FINDING MARS

You should begin this activity at sunset. An easy way to do this is to click the drop-down menu to the right of the **date & time** field on the **control panel**, and select **sunset**. Look toward the west by clicking the **W direction** button located on the **button bar** across the top of your screen, or by simply keying in the letter W (Mac users should refer to the button bar commands given at the beginning of this booklet). The screen will pan toward the west. Select a playing speed of **300×** normal time by clicking the drop-down menu at the right of the **time speed** field. Click the STOP **time mode** button when the Sun has set, the stars have come out, and dusk is almost over. Then click on the **constellations** button to show the constellations.

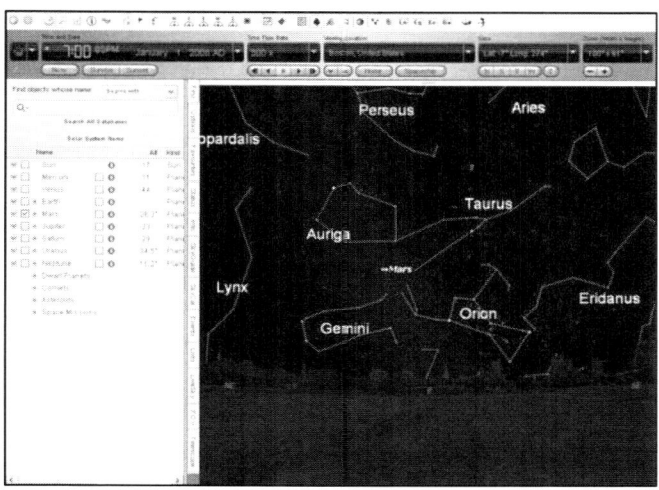

Click the FIND tab on the left **side pane**. A list of planets should appear. Those that are highlighted are currently up in your evening sky. Those that are not highlighted are not up in the sky at this time. We wish to find the planet Mars. If Mars is highlighted (see footnote if not highlighted)*, double-click or right click on it (control click for Mac) and select **center**. This will pan the screen and center Mars. You can now zoom in on Mars either by using the **zoom control** at the far right of the **control panel**, or by right clicking (control clicking for Mac) on the highlighted listing in the FIND **side pane**, and selecting **magnify**.

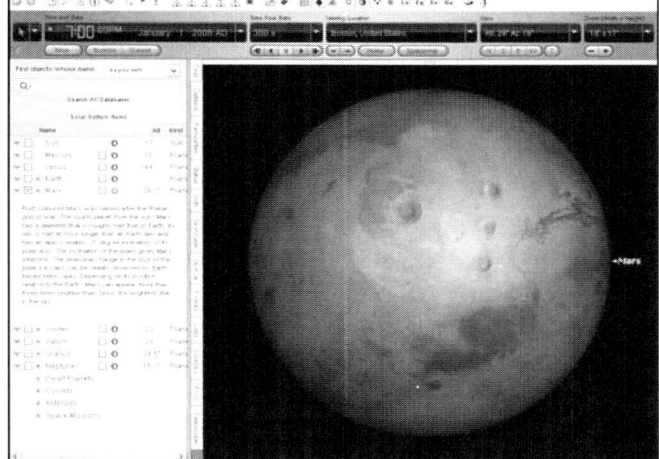

*If Mars is not currently highlighted, you will need to move time forward to a time when Mars will rise in the east. Start by looking toward the east by clicking the **E viewing direction** button located on the **button bar** across the top of your screen. Select a playing speed of **300×** or **3000×** normal time by clicking the drop-down menu at the right of the **time speed** field. Click the STOP **time mode** button when the planet Mars becomes highlighted in the FIND **side pane** listing. It may be that Mars is up during the day. If the Sun rises before Mars does, then click the **daylight** button on the **button bar**. This will keep the sky dark so you can see the stars and constellations. Once Mars has risen, double-click on it. This will pan the screen and center Mars. You can now zoom in on Mars either by using the **zoom control** at the far right of the **control panel**, or by right clicking (control clicking for Mac) on the highlighted listing in the FIND **side pane** and selecting **magnify**.

Click the information icon **(i)** next to Mars in the **side pane** to read a short description of the planet. For even more information, select the INFO tab on the left **side pane** and click the plus sign (gray arrow for Mac) to expand the different information categories.

PART 2: THE MARTIAN MOONS PHOBOS & DEIMOS

We wish to take a closer look at Mars's two moons, Phobos and Deimos, both thought to be captured asteroids.

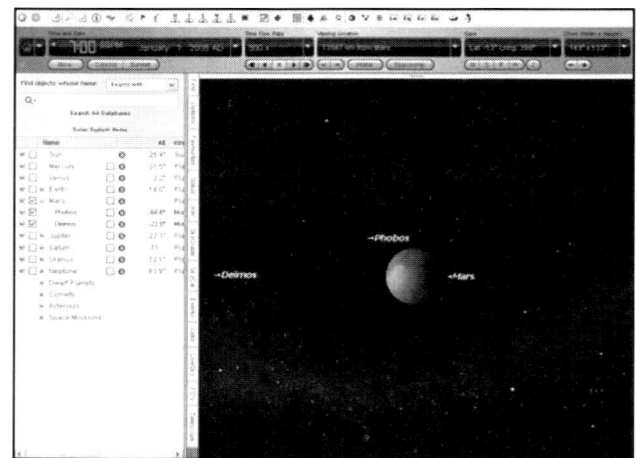

Click the plus sign (gray arrow for Mac) next to **Mars** in the FIND **side pane**. This will list Mars's two moons: Phobos and Deimos. Click the first check box for both moons. These moons are now labeled on your screen. In order to see both of them, you may need to pan out by using the **zoom control** at the far right of the **control panel**. Keep in mind that the moons may be out of sight behind the planet itself. In that case, use the **3000× time speed** setting on the **control panel** to move time forward so that both moons are clearly visible. Remember that as you move time forward, you may need to click the **horizon** and **daylight** buttons on the **button bar** as needed.

Let's take a closer look at each moon. Double click on Phobos. This will center Phobos on your screen. Use the **zoom control** to get a close-up view.

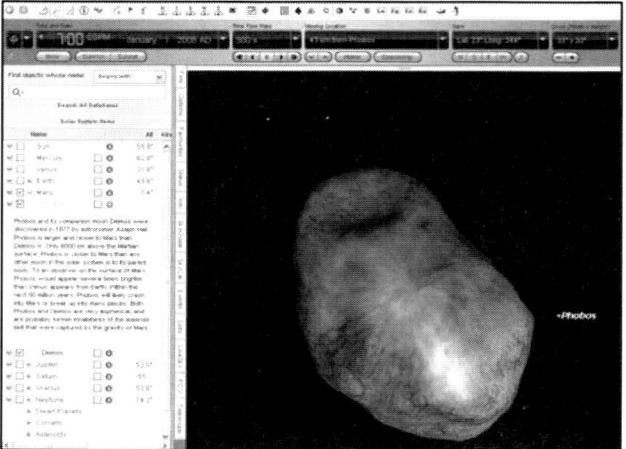

[OPTIONAL] If you would like to get an even closer view beyond the limits of the zoom function, right click (control click for Mac) on the moon and select **go there**. You are now on the surface (to see the surface, make sure your horizon is turned on). Now launch yourself up above the moon by clicking the **location above surface of planet** up arrow on the **control panel**. Lock onto Phobos by double-clicking on **Phobos** in the FIND **side pane**, then use the **zoom control** as needed to see the moon full screen. When done, be sure to click the **back** button on the **button bar** a few times to return to an earthbound perspective for the rest of the assignment.

Click the information icon **(i)** in the **side pane** to read a short description of this moon, then select the INFO tab on the left **side pane** and click the plus sign (gray arrow for Mac) to expand the different information categories.

Observe Deimos as described above and complete the following table for both moons:

Compare to Earth's Moon with radius = 1737 km and mass = 7.35×10^{22} kg = 0.0123 Earth masses.
Mars's radius is 3396 km, and Mars' mass is 0.1073 Earth masses. Earth's mass = 5.974×10^{24} kg.

ATTRIBUTE	PHOBOS	DEIMOS
Radius (in km):		
…in multiples of Earth's Moon radius:*		
Mass (in kg):		
…in Earth masses:		
…in multiples of Earth's Moon mass:		
…as ratio of Mars's mass:		
Orbit size (in km):		
…in multiples of Mars's radius:		
Sidereal day (one rotation):		
Solar day (noon to noon):		
Year (to orbit around Mars):		

*The info given for *radius* may be in km. To compute the *multiple of Earth's Moon radius*, take *radius* in km and divide by *Earth's Moon radius* as given above. Follow the same procedure for *mass* and *orbit size*.

Note how small the ratio of each moon's mass is as compared to the mass of Mars itself. Although Mars is not very large, the moons are so small that their mass ratio is small, with little gravitational effect on Mars. In contrast, Earth's moon-to-planet mass ratio is one of the largest in the solar system at 0.0123 (or 1.23%).

PART 3: SATELLITE ORBITAL CHARACTERISTICS

Zoom back out so you can see Mars and its moons once again. Double-click on **Mars** in the FIND **side pane** to center on Mars and zoom in and out so you can see both moons. Select the **3000× time speed** on the **control panel** and watch the moons orbit Mars. You may need to click the **horizon** and **daylight** buttons on the **button bar** as needed. Watch the motion for a bit, adjusting the **zoom control** as needed. The tilting is due to our local perspective. You can remove this effect by selecting **view/ecliptic guides**, then **the ecliptic**. Why do the moons move from side to side rather than in circles around Mars? This is due in part to perspective. Since Mars's satellite orbital plane lies along the ecliptic, the plane of our solar system, we can only observe Mars from the side.

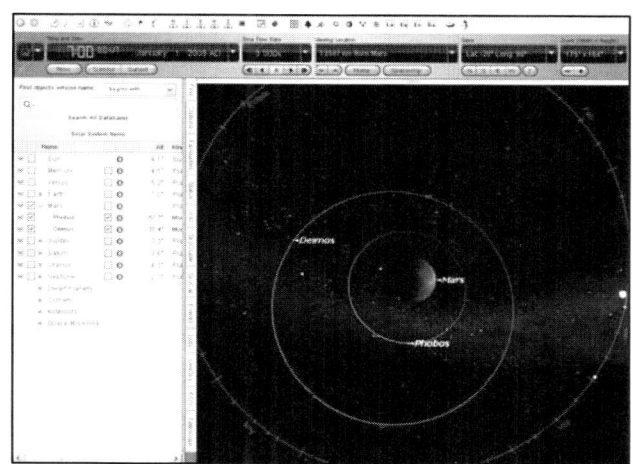

[OPTIONAL] You can use Starry Night™ Pro to observe the moons from above Mars's polar region. First use Starry Night™ Pro to go to Mars, then change your position to Mars's north pole. From there, launch yourself straight up until you can view the moons from above. Now, when you click the PLAY button, you will see the moons orbiting in nearly circular orbits. When done, click the **back** button repeatedly until you have returned to your original perspective from Earth.

To better visualize the satellite orbital motion, click both the first and second set of check boxes in the FIND **side pane** for both moons to label and trace out their respective orbits around Mars. Again, you may need to adjust the **zoom controls** to better view the orbital plane. Although they are too small to observe through amateur telescopes, Phobos and Deimos move quickly around Mars and can significantly change their position in a matter of hours.

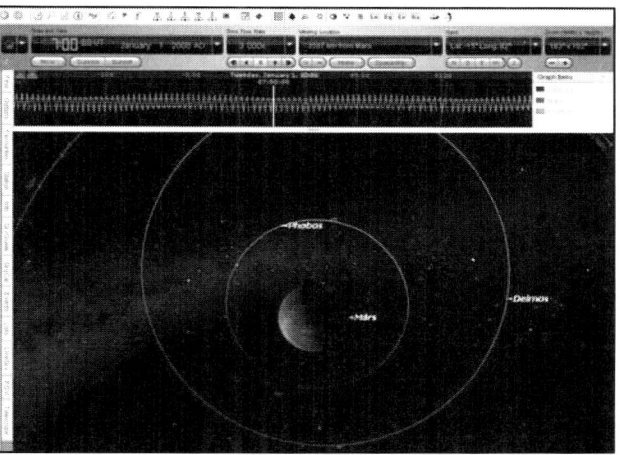

Select **now** from the **date & time** display on the **control panel**. Now slowly move time forward using the **300×** or **3000× time speeds** and observe the orbital motion as it would appear at this date and time as seen from Earth.

Which moon is traveling faster? _____.
Which moon is traveling slower? _____.

To see some different viewing perspectives of the moons, right click (control click for Mac) on Mars and select **graph elongation of moons**. The graph shows the angular separation between the moons with Mars. Notice the high frequency for Phobos. You can move time forward with the **time speed** fields on the **control panel**, or you can simply grab the graph by clicking down on it with the mouse then dragging to the left or right while holding the mouse button down. You can also expand the horizontal resolution of the graph by clicking on the plus and minus indicators at the top left of the graph. To increase the vertical resolution, grab the bottom of the graph with the mouse and pull down. By moving the graph back and forth to look for different alignments, you can answer the following questions.

SATELLITE ORBITAL CHARACTERISTICS:

Time for a complete cycle of Phobos:	
Time for a complete cycle of Deimos:	
Maximum elongation for Phobos (see vertical scale)*:	
Maximum elongation for Deimos (see vertical scale)*:	
First viewing opportunity for both moons on the same side of Mars**:	
First viewing opportunity for both moons on the opposing side of Mars:	

*You can check and improve on the precision of the maximum elongation values by centering your pointer on Mars (taking care to hover your pointer right over the center of Mars until an arrow appears), then clicking down and dragging the pointer to the farthest extent of the moon's orbit (make sure the orbit trace is on). The angular separation will be displayed.

**To find the first viewing opportunity for both moons on the same side of Mars, you will need to expand the horizontal axis as described above. Then move time forward, or grab the graph and advance time to the first time that both lines are above (or below) the central line (which represents the planet Mars). Don't forget to check and see if this is during our daytime or nighttime. You can check either by noting the Starry Night™ Pro time display or by noting the day and night shading on the graph after expanding the horizontal resolution. Your times above should indicate nighttime observing times.

Activity 09—THE PLANET JUPITER

*These activities are designed to work with the Starry Night software that comes with your text, from any home location you choose, and with the current date and time, unless indicated otherwise. You may always revert to factory default settings by clicking **FILE/ preferences**, then selecting **factory defaults** as needed. You may also undo a command or series of commands on the PC by clicking the **back** button at the top left of the **button bar**. You should refer to the key given at the beginning of this booklet for clarification of "on screen" buttons, controls, and functions. PC **button bar** items can all be accessed through the **menu**. "Right click" on the PC is equivalent to "control click" on the Mac. All activities assume that OpenGL graphics capabilities are enabled on your computer.*

PART 1: FINDING JUPITER

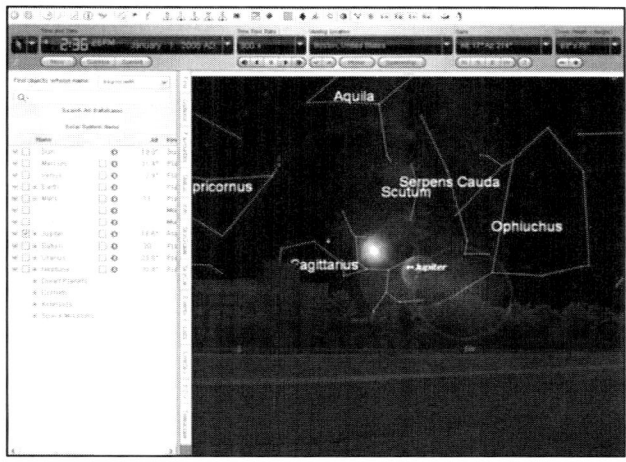

You should begin this activity at sunset. An easy way to do this is to click the drop-down arrow to the right of the **date & time** field on the **control panel**, and select **sunset**. Look toward the west by clicking the **W viewing direction** button located on the **button bar** across the top of your screen, or by simply keying in the letter W (Mac users should refer to the button bar commands given at the beginning of this booklet). The screen will pan toward the west. Select a playing speed of **300×** normal time by clicking the drop-down menu at the right of the **time speed** field. Click the STOP **time mode** button when the Sun has set, the stars have come out, and dusk is almost over. Then click on the **constellations** button to show the constellations.

Click the FIND tab on the left **side pane**. A list of planets should appear. Those that are highlighted are currently up in your evening sky. Those that are not highlighted are not up in the sky at this time. We wish to find the planet Jupiter. If Jupiter is highlighted (see footnote if not highlighted)*, double-click or right click on it (control click for Mac) and select **centre**. This will pan the screen and center Jupiter. You can now zoom in on Jupiter either by using the **zoom control** at the far right of the **control panel**, or by right clicking (control clicking for Mac) on the highlighted listing in the FIND **side pane** and selecting **magnify**.

*If Jupiter is not currently highlighted, you will need to move time forward to a time when Jupiter will rise in the east. Start by looking toward the east by clicking the **E viewing direction** button located on the **button bar** across the top of your screen. Select a playing speed of **300×** or **3000×** normal time by clicking the drop-down menu at the right of the **time speed** field. Click the STOP **time mode** button when the planet Jupiter becomes highlighted in the FIND **side pane** listing. It may be that Jupiter is up during the day. If the Sun rises before Jupiter does, then click the **daylight** button on the **button bar**. This will keep the sky dark so you can see the stars and constellations. Once Jupiter has risen, double-click on it. This will pan the screen and center Jupiter. You can now zoom in on Jupiter either by using the **zoom control** at the far right of the **control panel**, or by right clicking (control clicking for Mac) on the highlighted listing in the FIND **side pane** and selecting **magnify**.

Click the information icon **(i)** in the **side pane** to read a short description of the planet Jupiter. Next, select the INFO tab on the left **side pane** and click the plus sign (gray arrow for Mac) to expand the different information categories. Use the information in the OTHER DATA section to complete the following table of Jupiter's physical characteristics and compare to Earth.

(You will need to know Earth's radius = 6378 km. To compute percent, divide one value by the other and multiply by 100.)

PHYSICAL CHARACTERISTICS:

Radius of Jupiter:	_____km, _____% of Earth
Mass of Jupiter:	
Angular size in arc minutes and seconds as seen from Earth:	
Orbit size (mean distance from the Sun):	
Sidereal day (period of rotation):	
Solar day (noon to noon on surface):	
Length of year:	

Complete the following table of Jupiter's observational characteristics:*

OBSERVATIONAL CHARACTERISTICS:

Date, time	
Azimuth, altitude	
Currently in which constellation:	
Current apparent magnitude:	
Max possible magnitude as seen from Earth:	
Current distance from Earth:	
Distance at conjunction (aligned with the Sun):	
Distance at opposition (opposite the Sun):	

*For *constellation, azimuth, and altitude,* see POSITION IN SKY; for *distance from Earth,* see POSITION IN SPACE. For everything else, see OTHER DATA. To compute *distance at conjunction,* take *orbit size* and add Earth's distance from the Sun of 1 AU. To compute *distance at opposition,* take *orbit size* and subtract Earth's distance from Sun of 1 AU.

PART 2: JUPITER'S ORBITAL CHARACTERISTICS

Use the **zoom control** at the far right of the **control panel** to zoom back out to full-scale view. Right click (control click for Mac) on Jupiter and select **orbit**. This shows Jupiter's orbital path as seen from Earth. You will need to click the **horizon** button on the **button bar** to get rid of the Earth's horizon should it interfere with your view of Jupiter's orbit. You should also click the **daylight** button to turn off daylight. You should still be locked on to Jupiter. If not, right click (control click for Mac) on the planet Jupiter, and then select **center**. To maintain the proper perspective, select **view/ecliptic guides**, then **the ecliptic**. Next, select a **time speed** of **1 day** and click the PLAY **time mode** button. If you need to slow down or speed up, adjust the **time speed** fields as needed. You should be able to see Jupiter

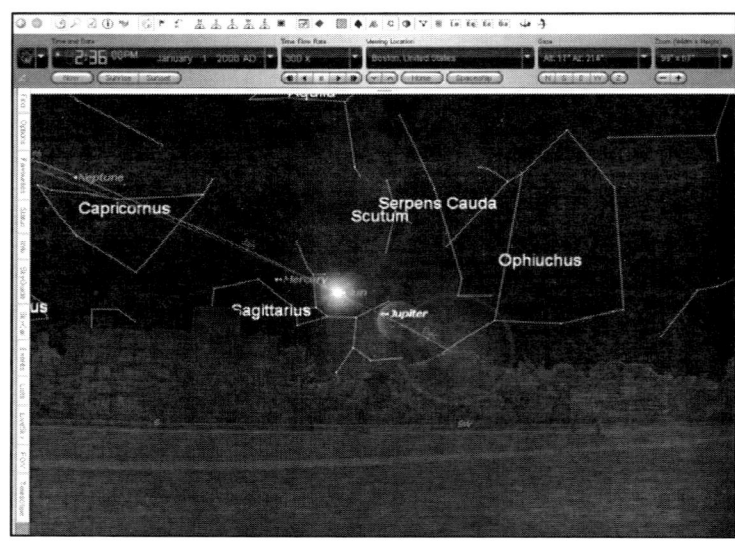

locked in the center of your field of view, but moving across the star background, and you should be able to complete a full revolution in less than a minute.

For a superior planet such as Jupiter, it is of interest to determine the dates of *conjunction* and *opposition*. *Conjunction* is when the planet lines up with the Sun as viewed from Earth. This will occur when the planet is on the opposite side of the Sun as compared to us on Earth. *Opposition* is when the planet is on the same side of the Sun as us, but beyond the Earth so that it appears high in our night sky around midnight. This is the best position for observing the planet (being high in the night sky and positioned closest to the Earth due to its orbit).

To determine the date of *conjunction*, simply step through the orbit using the **time mode** and **time speed** buttons on the **control panel** until the planet lines up with the Sun (it won't be exact, as it usually passes just above or below the Sun). You may find it helpful to step forward and backward one day at a time until you have the best alignment. Use this method to find the next date of conjunction for Jupiter.

Date of Jupiter at next conjunction:

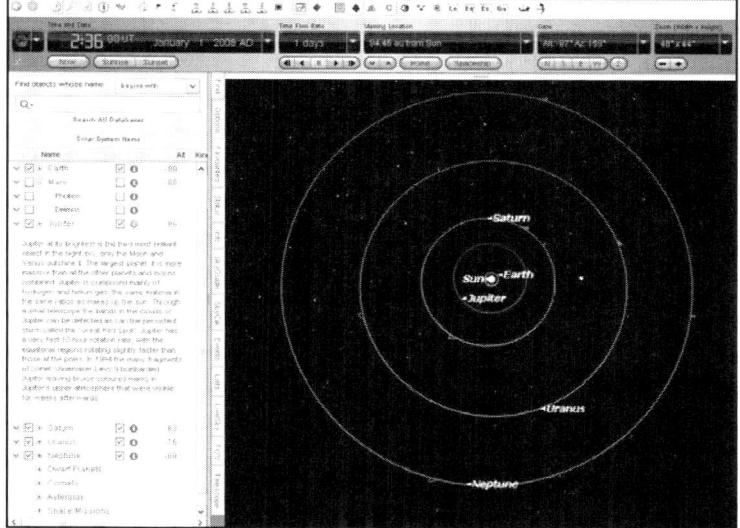

Now for the big picture: To view the position of Jupiter as seen from outer space and compared to the positions of the other planets, select **favourites/solar system/outer planets/** and select **outer solar system**. Click the STOP **time mode** button, select **now** from the **date and time** drop-down menu, then use the **zoom control** at the upper right of the **button bar** to zoom in so that Jupiter's orbit fills the screen. Click both the planet and orbit boxes for Earth in the FIND **side pane** to show Earth's orbit.

Now draw the relative positions of Earth and Jupiter as you see them for today's date in the space below:

TODAY'S DATE:

Let's view the conjunction event from this solar system perspective. Advance time forward using the PLAY **time mode** button, then STOP and FORWARD STEP and BACK STEP until you are at the conjunction date you found earlier. Draw the relative positions of the Earth and Jupiter as you see them during this conjunction event in the following space.

DATE OF CONJUNCTION:

To determine the date of opposition, step forward one day at a time until the Earth is right in between the Sun and Jupiter. Try and imagine a straight line from the Sun, through Earth, to Jupiter. This method is not very precise, but will give you a general idea of when opposition will occur.*

Date of Jupiter at next opposition:

Draw the relative positions of the planets as you see them during this opposition event in the space below.

DATE OF OPPOSITION:

*To improve on your estimate of Jupiter at opposition, experiment with your **viewing location** fields to better position yourself so you can best see the opposition event. Click the **location** button on the **button bar** or, from the **menu**, experiment with **options/viewing location**. Select Sun, set your latitude for 90 degrees, and then use the **location above surface of planet** arrows on the **control panel** to launch yourself straight up until you have a face-on view of the solar system. You should turn on the orbits for Earth and Jupiter by clicking the FIND **side pane** and checking both boxes for each planet. First play, and then step forward and backward one day at a time to get the best possible alignment for opposition. Turning on the LOCAL (lo) **coordinate system** on the **button bar** can help with the alignment. As you move time forward, to keep from rotating with the Sun, you should select **follow Sun in orbit** from the **location control** drop-down menu on the **control panel**.

When done, click the **back** button repeatedly to return to Earth with your previous software configuration, or you can manually turn off planet labels and orbital tracks using the FIND **side pane**.

PART 3: FLYING TO JUPITER

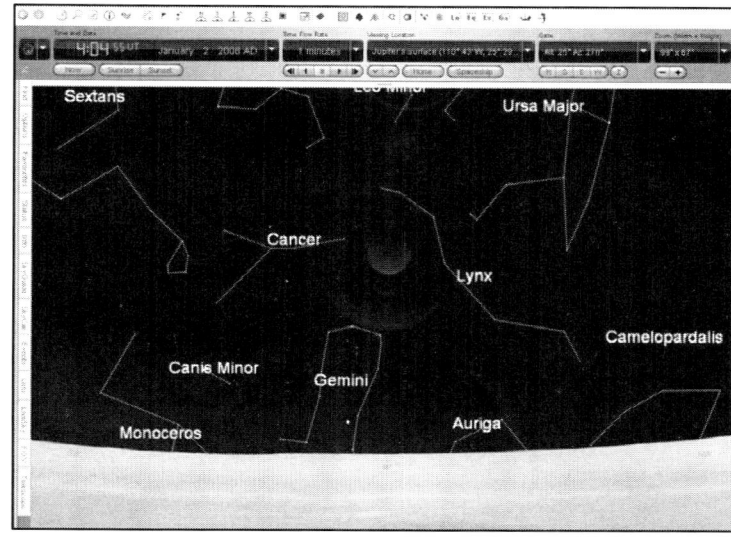

Let's see what a day would be like on Jupiter (technically referred to as a *solar day*). The easiest way to experience a solar day is to watch a sunset, take note of the date and time, then watch another consecutive sunset, and see what the time difference is between them.*

Right click (control click for Mac) on Jupiter and select **go there**. You can animate the journey by first unchecking the **only animate intra planet changes** box under **file/preferences/ responsiveness** (also, be sure to have your horizon turned on to see the photorealistic surface panorama). Click the **W viewing direction** button on the **button bar**, or simply hit the W key on the keyboard. Select a **time speed** of **1 minute**. Click the PLAY **time mode** button, and click STOP when the Sun is near setting.** Use the FORWARD STEP and BACKWARD STEP **time mode** buttons until you see the Sun just starting to set. You may then need to select a smaller unit of time until the Sun is just touching the horizon (Jupiter has no solid surface, so it's really the cloud tops we are viewing). Note the date and time, then continue on to the next sunset and calculate the time difference between them.

Date and time of first sunset on Jupiter:	
Date and time of second sunset on Jupiter:	
Subtract to get length of solar day:	

*For even greater accuracy, instead of using sunset, you should use solar noon. To find solar noon, use the LOCAL (lo) **coordinate system** button to align the Sun with the local meridian line, the Sun's highest point in the sky for that day. Starry Night™ Pro also gives the time of solar noon in the **date & time** drop-down menu on the **control panel**.

If you are unable to see the Sun to the west during sunset, first use the **date & time drop-down menu to select **sunset**, then click the check box to the left of the Sun in the FIND **side pane** to label the Sun, then pan to the left (SW) or right (NW) until you see it. If you still have trouble locating the Sun, right click (control click for Mac) on the Sun in the FIND **side pane** and select **centre**. You will need to unlock the Sun before continuing with this activity. The easiest way to "unlock" is to simply "grab" the sky by clicking and holding the left mouse button and then moving the mouse a little bit in any direction.

The solar day is not the same as the sidereal rotation period of a planet. This is because the solar day takes into account both the rotation of the planet and the revolution of the planet around the Sun. Your calculation of a solar day is, therefore, a little different than the sidereal day found earlier in this activity from the INFO **side pane**.

How do we determine the revolutionary period, or a planet's "year?"

Return to the solar system view. From the **menu**, select **favourites/solar system/outer planets/** and select **outer solar system**. Click the STOP **time mode** button, then select **now** from the **date and time** drop-down menu. Choose a **time speed** of **1 day**, click the PLAY **time mode** button, then note the time it takes for the planet to return to its original position. You may find it easiest to align the planet to the farthest left or right before starting. Complete the chart below.

Date of Jupiter at start:	
Date of Jupiter after one revolution:	
Subtract to get length of year:	

Activity 10—JUPITER'S MOONS

*These activities are designed to work with the Starry Night software that comes with your text, from any home location you choose, and with the current date and time, unless indicated otherwise. You may always revert to factory default settings by clicking **FILE/ preferences**, then selecting **factory defaults** as needed. You may also undo a command or series of commands on the PC by clicking the **back** button at the top left of the **button bar**. You should refer to the key given at the beginning of this booklet for clarification of "on screen" buttons, controls, and functions. PC **button bar** items can all be accessed through the **menu**. "Right click" on the PC is equivalent to "control click" on the Mac. All activities assume that OpenGL graphics capabilities are enabled on your computer.*

PART 1: FINDING JUPITER

You should begin this activity at sunset. An easy way to do this is to click the drop-down menu to the right of the **date & time** field on the **control panel**, and select **sunset**. Look toward the west by clicking the **W viewing direction** button located on the **button bar** across the top of your screen, or by simply keying in the letter W (Mac users should refer to the button bar commands given at the beginning of this booklet). The screen will pan toward the west. Select a playing speed of **300×** normal time by clicking the drop-down menu at the right of the **time speed** field. Click the STOP **time mode** button when the Sun has set, the stars have come out, and dusk is almost over. Then click on the **constellations** button to show the constellations.

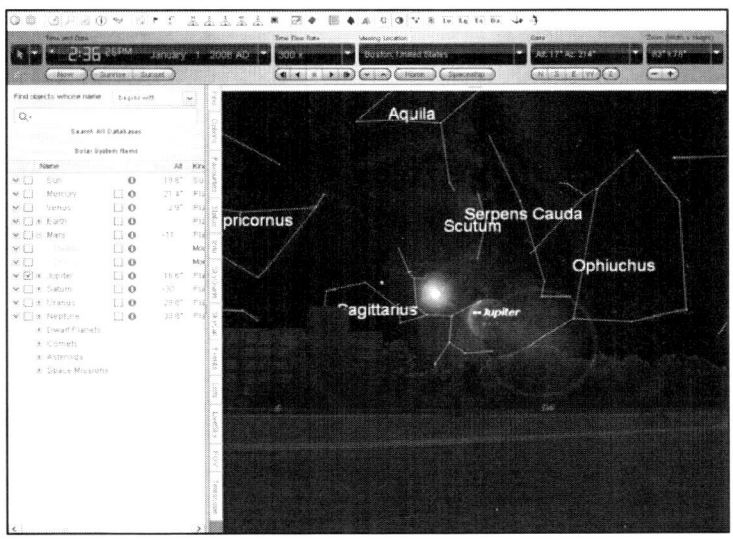

Click the FIND tab on the left **side pane**. A list of planets should appear. Those that are highlighted are currently up in your evening sky. Those that are not highlighted are not up in the sky at this time. We wish to find the planet Jupiter. If Jupiter is highlighted (see footnote if not highlighted)*, double-click or right click on it (control click for Mac) and select **center**. This will pan the screen and center Jupiter. You can now zoom in on Jupiter either by using the **zoom control** at the far right of the **control panel**, or by right clicking (control clicking for Mac) on the highlighted listing in the FIND **side pane**, and selecting **magnify**.

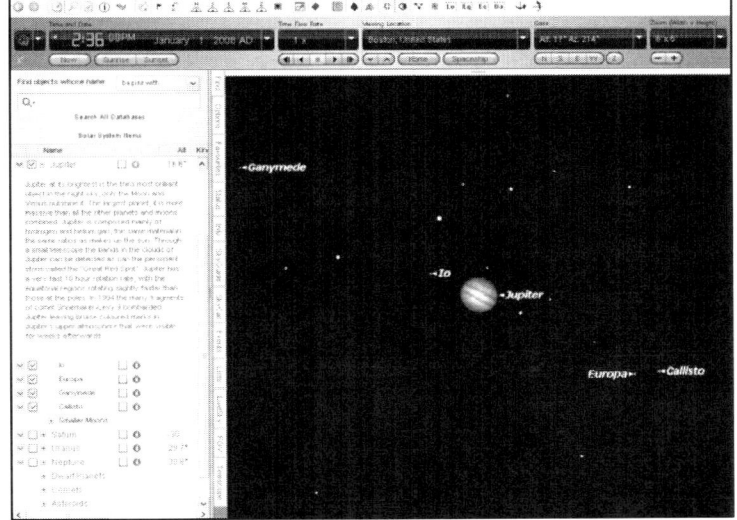

*If Jupiter is not currently highlighted, you will need to move time forward to a time when Jupiter will rise in the east. Start by looking toward the east by clicking the **E viewing direction** button located on the **button bar** across the top of your screen. Select a playing speed of 300× or 3000× normal time by clicking the drop-down menu at the right of the **time speed** field. Click the STOP **time mode** button when the planet Jupiter becomes highlighted in the FIND **side pane** listing. It may be that Jupiter is up during the day. If the Sun rises before Jupiter does, then click the **daylight** button on the **button bar**. This will keep the sky dark so you can see the stars and constellations. Once Jupiter has risen, double-click on it. This will pan the screen and center Jupiter. You can now zoom in on Jupiter either by using the **zoom control** at the far right of the **control panel**, or by right clicking (control clicking for Mac) on the highlighted listing in the FIND **side pane** and selecting **magnify**.

Click the information icon **(i)** in the **side pane** to read a short description of the planet Jupiter. For even more information, select the INFO tab on the left **side pane** and click the plus sign (gray arrow for Mac) to expand the different information categories.

PART 2: THE GALILEAN SATELLITES

Starry Night™ Pro will display 39 of Jupiter's moons. Keep in mind, though, that there are actually more than 60, and probably more to be discovered. We wish to take a closer look at the four largest moons, the Galilean satellites, first viewed by Galileo in 1610—Io, Europa, Ganymede, and Callisto.

Click the plus sign (gray arrow for Mac) next to JUPITER in the FIND **side pane**. This will list Jupiter's moons. Click the first check box for each of the four Galilean satellites—Io, Europa, Ganymede, and Callisto. These moons are now labeled on your screen. In order to see all four of them, you may need to pan out by using the **zoom control** at the far right of the **control panel**. Keep in mind that the moons may be out of sight behind the planet itself. In that case, use the **3000× time speed** setting on the **control panel** to move time forward so that all four moons are clearly visible. Remember that as you move time forward, you may need to click the **horizon** and **daylight** buttons on the **button bar** as needed.

Let's take a closer look at each moon. Double-click on **Io**. This will center Io on your screen. Use the **zoom control** to get a close-up view.

Click the information icon **(i)** in the **side pane** to read a short description of this moon. Then, select the INFO tab on the left **side pane** and click the plus sign (gray arrow for Mac) to expand the different information categories.

Observe each of the four Galilean Satellites as described above and complete the following table.
Compare to Earth's Moon with radius = 1737 km, and mass = 7.35×10^{22} kg = 0.0123 Earth masses. 1 AU = 1.50×10^8 km. Jupiter's radius is 71491 km and Jupiter's mass is 317.57 Earth masses. Earth's mass = 5.974×10^{24} kg.

ATTRIBUTE	IO	EUROPA	GANYMEDE	CALLISTO
Radius (in km):				
…in multiples of Earth's Moon radius:*				
Mass (in Earth Masses):				
…in multiples of Earth's Moon mass:				
…as ratio of Jupiter's mass:				
Orbit Size (in AU):				
…in km:				
…in multiples of Jupiter's radius:				
Sidereal day (one rotation):				
Solar day (noon to noon):				
Year (to orbit around Jupiter):				

*The info given for *radius* may be in km. To compute the *multiple of Earth's Moon radius*, take *radius* in km and divide by *Earth's Moon radius* as given above. Follow the same procedure for *mass* and *orbit size* as applicable.

Note how small the ratio of each moon's mass is compared to the mass of Jupiter itself. Although the Galilean satellites are some of the largest moons in the solar system, their mass ratio is very small, with little gravitational effect on Jupiter. In contrast, Earth's moon-to-planet-mass ratio is one of the largest in the solar system at 0.0123 (or 1.23%).

Note how close Io is to its parent planet in multiples of Jupiter's radius. Compare this with our own moon, whose distance from Earth is more than 60 times Earth's radius. When a large moon is too close to its parent planet, the tidal forces are so great that they try to rip the moon apart. The critical distance, known as the Roche limit, is the closest distance a large moon can exist in orbit around a planet without being ripped to shreds. Moons are thus generally found outside the Roche limit, and ring systems are usually found within the Roche limit. An interesting point to make is that although Io is outside the Roche limit, it is relatively close and is, therefore, subjected to tremendous tidal forces. These forces cause internal friction and generate heat. Thus Io has a hot interior and is volcanically active. Use the information provided on Io in the **description** section of the INFO **side pane**—or click the information icon **(i)** in the FIND **side pane** to answer the following questions.

Percentage of volcanoes covering Io's surface:	
Number of active volcanoes found by *Voyager 1*:	
Average surface temperature of Io:	
Temperature near active volcanoes:	
Maximum height of volcanic eruption plumes:	

Jupiter's Galilean satellites are very interesting planetary bodies. Due to their size and proximity to Jupiter, the largest planet in our solar system, they exhibit a great range of physical properties and behave more like planets than they do moons. Io, for example, is considered the most volcanically active object in the solar system. Europa, due to a possible liquid ocean of water beneath a thin crust of ice, has a relatively smooth surfaces with low relief. Due to the presence of liquid water, Europa is considered one of the most likely places to find life in the solar system. Ganymede is the largest moon in our solar system, and Callisto has one of the most heavily cratered surfaces in the solar system, indicating an extremely old surface. Scientists have much more to learn about Jupiter's moons. These are not dead worlds, with Io and Europa being highly active planetary objects that exhibit many interesting properties and, in the case of Europa, the potential for harboring life.

PART 3: SATELLITE ORBITAL CHARACTERISTICS

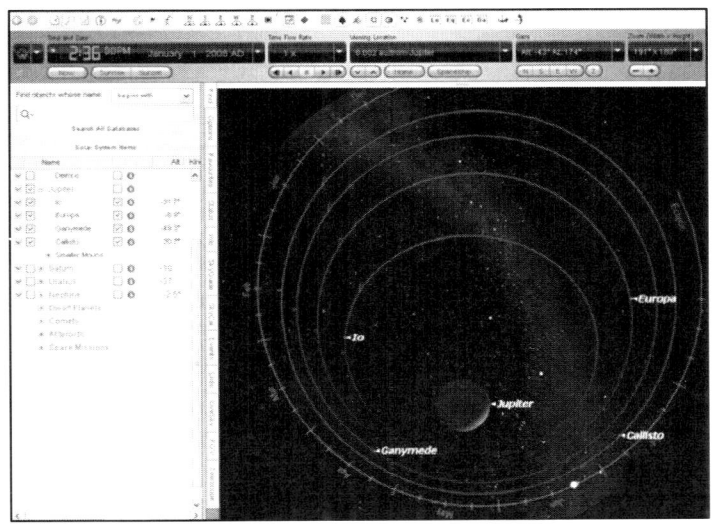

Zoom back out so you can see Jupiter and its moons once again. Click the **labels** button on the **button bar** to see the names of other moons. Double-click on **Jupiter** in the FIND **side pane** to center on Jupiter, and zoom in and out a bit more to see how many moons you can observe. Select the **3000× time speed** on the **control panel** and watch the moons orbit Jupiter. Click the **daylight** and **horizon** buttons as needed. For best results, so that the smaller moons will be displayed, Jupiter should cover about a third of your screen. Watch the motion for a bit. The tilting is due to our local perspective. You can remove this effect by selecting **view/ecliptic guides**, then **the ecliptic**. Why do the moons move from side to side rather than in circles around Jupiter? This also has to do with perspective. Since Jupiter's satellite orbital plane lies along the ecliptic, the plane of our solar system, we can only observe Jupiter from the side.

[OPTIONAL] You can use Starry Night to observe the moons from above Jupiter's polar region. First, use Starry Night™ Pro to go to Jupiter, then change your position to Jupiter's north pole. Next, launch yourself straight up until you can view the moons from above. Now, when you click the **play** button, you will see the moons orbiting in nearly circular orbits. When done, click the **back** button repeatedly until you have returned to your original perspective from Earth.

To better visualize the moons' orbital motion, click both the first and second set of check boxes in the FIND **side pane** for each of the four Galilean satellites to label and trace their respective orbits around Jupiter. Again, you may need to adjust the **zoom controls** to better view the orbital plane. Turn off the labels by once again clicking the **labels** button on the **button bar**, and animate by selecting a **time speed** of **3000×**.

Which moon is traveling the fastest? _____.
Which moon is traveling the slowest? _____.

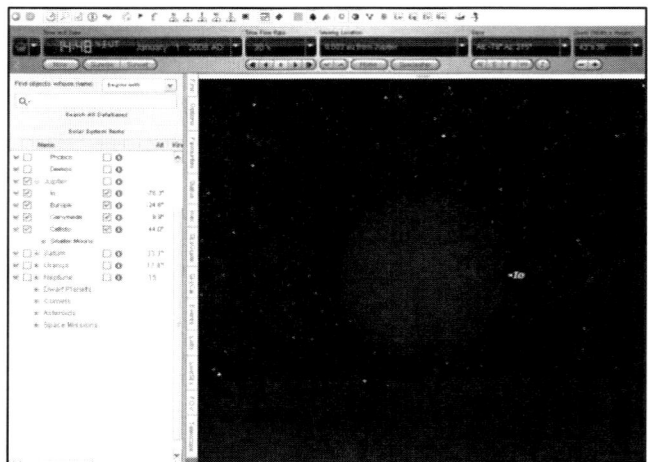

The fastest moon moves so quickly around Jupiter that we can often note a change in its position during a single evening's observing session. Select **now** from the **date & time** display on the **control panel**. Now slowly move time forward using the **300×** or **3000× time speeds** until just before the moon begins to pass behind the limb of Jupiter. Double-click on the moon to center it, then zoom in and observe this event slowly using the **30× time speed**. If this occurs on the dark side, you will have to take extra care with your observations. If need be, refer to the Starry Night™ Pro instruction manual to brighten the dark sides of planets and moons. Take note of the following times.

EVENT	DATE	TIME
1. Time moon begins passing behind Jupiter's limb:		
2. Time moon completes passing behind Jupiter's limb:		
3. Time moon begins to reemerge from behind Jupiter's limb:		
4. Time moon completely reemerges from behind Jupiter's limb:		

CALCULATIONS:

Calculate the time moon took to pass behind the limb (subtract line 1 from 2):	
Calculate the time moon took to transit behind Jupiter (subtract line 2 from 3):	
Calculate the time moon took to reemerge from behind the limb (subtract line 3 from 4):	

To see some different viewing perspectives of the moons, right click (control click for Mac) on Jupiter and select **graph elongation of moons**. The graph shows the angular separation between the four Galilean satellites with Jupiter. Notice the high frequency for our fastest moon. You can move time forward with the **time speed** fields on the **control panel**, or you can simply grab the graph by clicking down on it with the mouse then dragging to the left or right while holding the mouse button down. You can also expand the horizontal resolution of the graph by clicking on the plus and minus indicators at the top left of the graph.

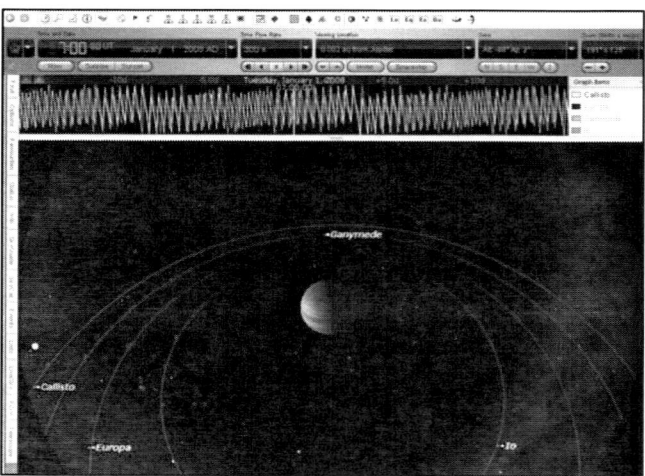

To increase the vertical resolution, grab the bottom of the graph with the mouse and pull down. By moving the graph back and forth to look for different alignments, you can answer the following questions.

GALILEAN SATELLITE ORBITAL CHARACTERISTICS:

Time for a complete cycle of Io:	
Time for a complete cycle of Europa:	
Time for a complete cycle of Ganymede:	
Time for a complete cycle of Callisto:	
Maximum elongation for Io (see vertical scale)*:	
Maximum elongation for Europa (see vertical scale)*:	
Maximum elongation for Ganymede (see vertical scale)*:	
Maximum elongation for Callisto (see vertical scale)*:	
First viewing opportunity for all the moons on the same side of Jupiter**:	
First viewing opportunity for all the moons on the opposing side of Jupiter:	

*You can check and improve on the precision of the maximum elongation values by centering your pointer on Jupiter (take care to hover your pointer right over the center of Jupiter until an arrow appears), then clicking down and dragging the pointer to the farthest extent of the moon's orbit (make sure the orbit trace is on). The angular separation will be displayed.

**To find the first viewing opportunity for all the moons on the same side of Jupiter, you will need to expand the horizontal axis as described above. Then move time forward, or grab the graph and advance time to the first time that all the lines are above (or below) the central line (which represents the planet Jupiter). Don't forget to check and see if this is during our daytime or nighttime. You can check either by noting the Starry Night™ Pro time display or by noting the day and night shading on the graph after expanding the horizontal resolution. You should determine nighttime observing times when answering the questions above.

PART 4: [OPTIONAL] KEPLER'S LAWS

You can use the information gathered in the Galilean Satellite Orbital Characteristics table (above) to test Kepler's third law of planetary motion. The third law relates a planet or satellite's orbital period, T, to its radial distance, R, (more technically, the semi-major axis). The relationship can be expressed as follows:

$$T^2 = [4\pi^2/GM]\ R^3,\ \text{where } G = 6.67 \times 10^{-11}\text{Nm}^2/\text{kg}^2,\text{ and } M \text{ is the mass of the planet}$$

Since we do not always know the mass of the planet (or indeed, we may want to determine the mass of the planet*), the best way to work with this equation is to graph T^2 as a function of R^3. If the relationship holds, we will get a straight line (linear relationship).

In this case, we don't have distance values for the radius, R, but we do have angular measures given by the maximum elongation values tabulated above. Due to the small angles involved, the angular values will correlate nicely with the radius.

First, tabulate R^3 and T^2 (use the values from the table above), then plot accordingly.

Moon	R^3	T^2

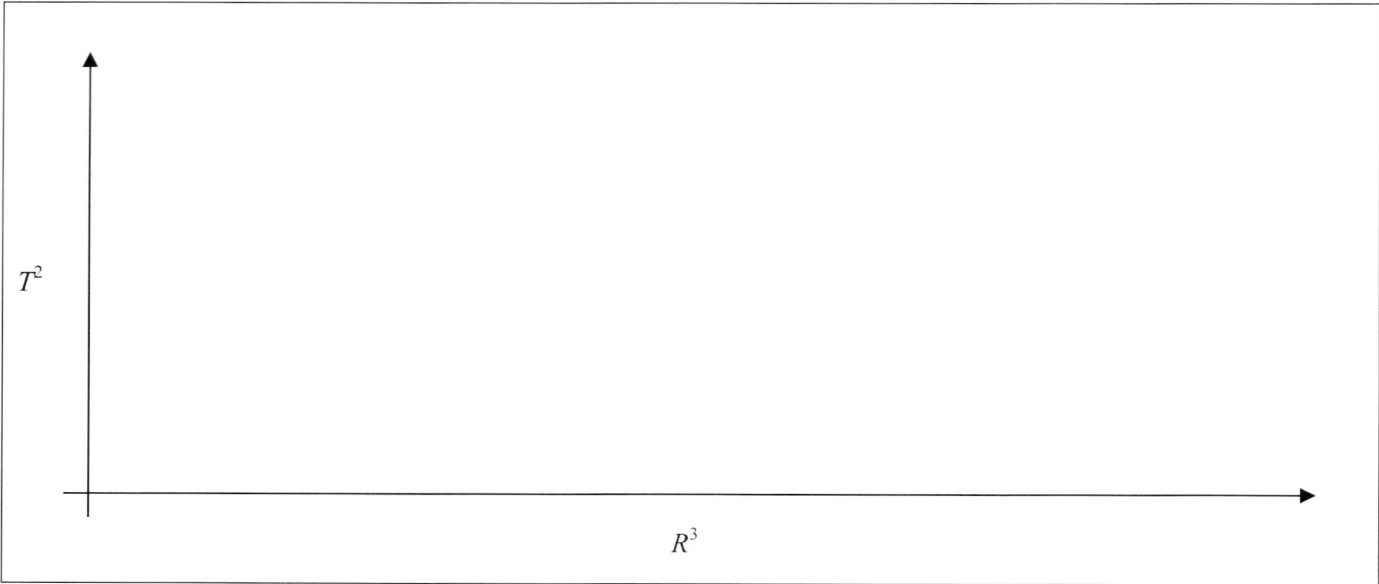

If you were careful with your data collection and calculations, you should see a clear linear relationship demonstrating the validity of Kepler's third law of planetary motion as applied to the moons of Jupiter. You may continue this exercise by plotting additional moons, and the linear relationship should be even clearer and better defined.

*If you were to relate the angular measure to radial distance in meters or kilometers, then you would find that the slope of the line is given by $[4\pi^2/GM]$. From this, you could calculate the mass of the planet Jupiter.

Activity 11—THE PLANET SATURN

*These activities are designed to work with the Starry Night software that comes with your text, from any home location you choose, and with the current date and time, unless indicated otherwise. You may always revert to factory default settings by clicking **FILE/ preferences**, then selecting **factory defaults** as needed. You may also undo a command or series of commands on the PC by clicking the **back** button at the top left of the **button bar**. You should refer to the key given at the beginning of this booklet for clarification of "on screen" buttons, controls, and functions. PC **button bar** items can all be accessed through the **menu**. "Right click" on the PC is equivalent to "control click" on the Mac. All activities assume that OpenGL graphics capabilities are enabled on your computer.*

PART 1: FINDING SATURN

You should begin this activity at sunset. An easy way to do this is to click the drop-down menu to the right of the **date & time** field on the **control panel** and select **sunset**. Look toward the west by clicking the **W viewing direction** button located on the **button bar** across the top of your screen, or by simply keying in the letter W (Mac users should refer to the button bar commands given at the beginning of this booklet). The screen will pan toward the west. Select a playing speed of **300×** normal time by clicking the drop-down menu at the right of the **time speed** field. Click the STOP **time mode** button when the Sun has set, the stars have come out, and dusk is almost over. Then click on the **constellations** button to show the constellations.

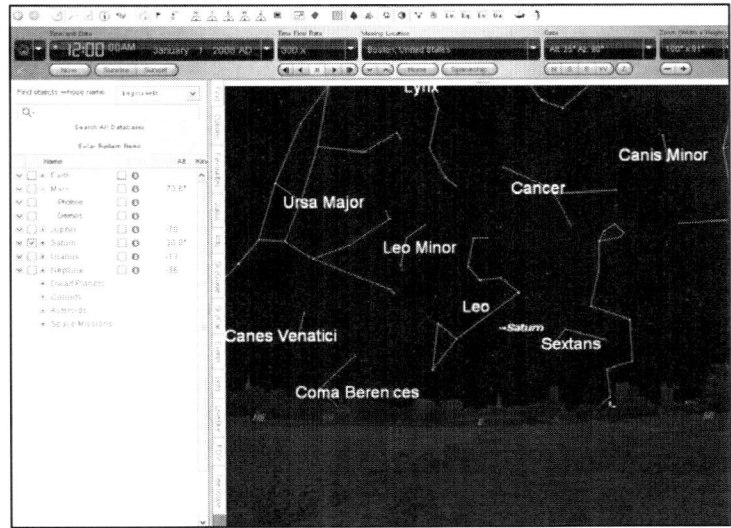

Click the FIND tab on the left **side pane**. A list of planets should appear. Those that are highlighted are currently up in your evening sky. Those that are not highlighted are not up in the sky at this time. We wish to find the planet Saturn. If Saturn is highlighted (see footnote if not highlighted)*, double-click or right click on it (control click for Mac) and select **center**. This will pan the screen and center Saturn. You can now zoom in on Saturn either by using the **zoom control** at the far right of the **control panel**, or by right clicking (control clicking for Mac) on the highlighted listing in the FIND **side pane** and selecting **magnify**.

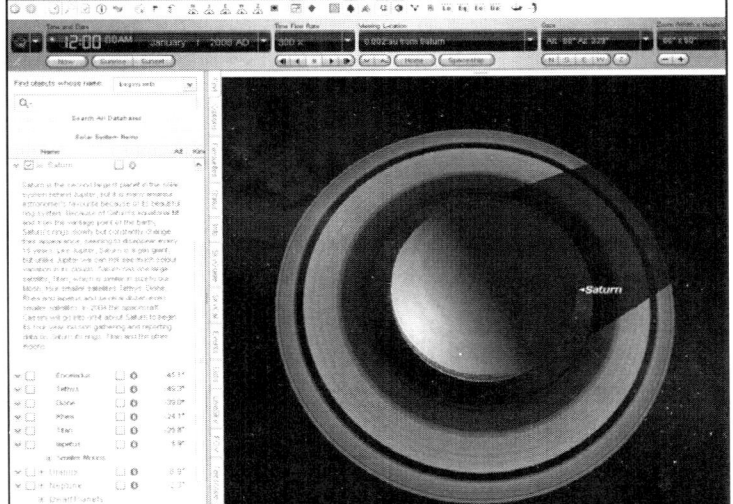

*If Saturn is not currently highlighted, you will need to move time forward to a time when Saturn will rise in the east. Start by looking toward the east by clicking the **E viewing direction** button located on the **button bar** across the top of your screen. Select a playing speed of **300×** or **3000×** normal time by clicking the drop-down menu at the right of the **time speed** field. Click the STOP **time mode** button when the planet Saturn becomes highlighted in the FIND **side pane** listing. It may be that Saturn is up during the day. If the Sun rises before Saturn does, then click the **daylight** button on the **button bar**. This will keep the sky dark so you can see the stars and constellations. Once Saturn has risen, double-click on it. This will pan the screen and center Saturn. You can now zoom in on Saturn either by using the **zoom control** at the far right of the **control panel**, or by right clicking (control clicking for Mac) on the highlighted listing in the FIND **side pane** and selecting **magnify**.

Click the information icon **(i)** in the **side pane** to read a short description of the planet Saturn. Next, select the INFO tab on the left **side pane** and click the plus sign (gray arrow for Mac) to expand the different information categories.

Use the information in the OTHER DATA section to complete the following table of Saturn's physical characteristics and compare to Earth.

(You will need to know Earth's radius = 6378 km. To compute percent, divide one value by the other and multiply by 100.)

PHYSICAL CHARACTERISTICS:

Radius of Saturn:	_____ km, _____ % of Earth
Mass of Saturn:	
Angular size in arc minutes and seconds as seen from Earth:	
Orbit size (mean distance from the Sun):	
Sidereal day (period of rotation):	
Solar day (noon to noon on surface):	
Length of year:	

Complete the following table of Saturn's observational characteristics:*

OBSERVATIONAL CHARACTERISTICS:

Date, time:	
Azimuth, altitude:	
Currently in which constellation:	
Current apparent magnitude:	
Max possible magnitude as seen from Earth:	
Current distance from Earth:	
Distance at conjunction (aligned with the Sun):	
Distance at opposition (opposite the Sun):	

*For *constellation, azimuth, and altitude,* see POSITION IN SKY; for *distance from Earth,* see POSITION IN SPACE. For everything else, see OTHER DATA. To compute *distance at conjunction,* take *orbit size* and add Earth's distance from the Sun of 1 AU. To compute *distance at opposition,* take *orbit size* and subtract Earth's distance from Sun of 1 AU.

PART 2: SATURN'S ORBITAL CHARACTERISTICS

Use the **zoom control** at the far right of the **control panel** to zoom back out to full-scale view. Right click (control click for Mac) on Saturn and select **orbit**. This shows Saturn's orbital path as seen from Earth. You will need to click the **horizon** button on the **button bar** to get rid of Earth's horizon should it interfere with your view of Saturn's orbit. You can click the **daylight** button to turn off daylight. You should still be locked on to Saturn. If not, right click (control click for Mac) on the planet Saturn, and then select **centre**. To maintain the proper perspective, select **view/ecliptic guides**, then **the ecliptic**.

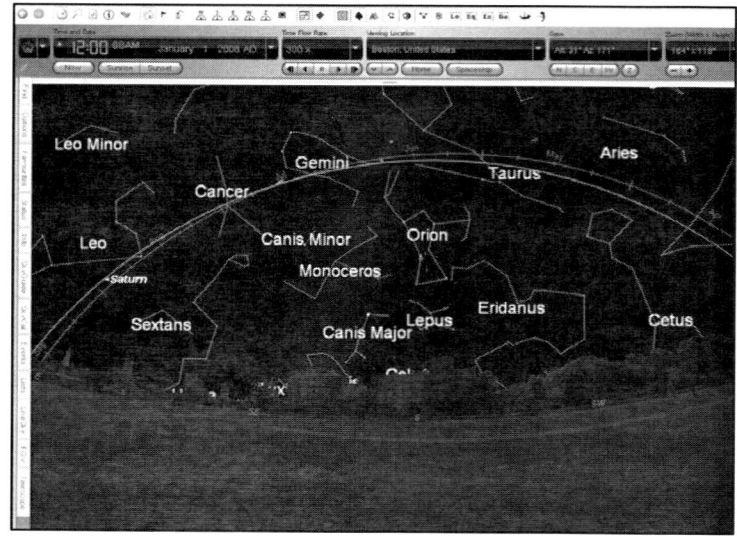

Start with today's date by entering it in the **date & time** field of the **control panel**, or by selecting **now** from the drop-down menu to the right of the **date & time** field. Note what constellation Saturn is in, then select a **time speed** of **1 day** and click the PLAY **time mode** button. If you need to slow down or speed up, adjust the **time speed** fields as needed.

You should be able to see Saturn locked in the center of your field of view, but moving across the star background. You should also be able to complete a full revolution in less than a minute. Click the STOP **time mode** button when Saturn has returned to the same constellation and note the date. Use this information to complete the following table.

VIEWING SATURN'S RELATIVE POSITION:

Constellation Saturn is located in today:	
Today's date (corresponding to location of Saturn given above):	
Date when Saturn is again located in same constellation:	
Length of time it takes Saturn to return to the same place in the sky (subtract the dates given above):	

*Note that since Saturn is relatively far from the Sun as compared to the Earth, this value will be close to Saturn's sidereal period (year). However, there will be a difference since we are looking at Saturn's position relative to the Earth, which is itself in an orbit around the Sun.

PART 3: FLYING TO SATURN

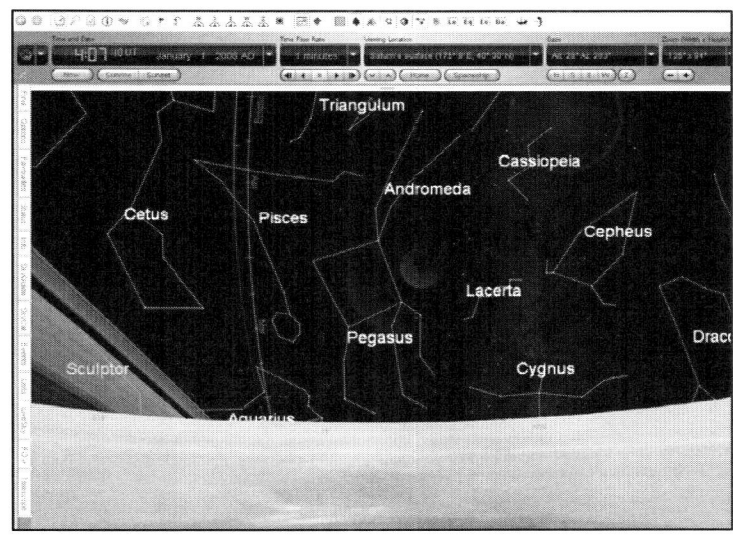

Let's see what a day would be like on Saturn (technically referred to as a *solar day*). The easiest way to experience a solar day is to watch a sunset, take note of the date and time, then watch another consecutive sunset, and see what the time difference is between them.*

Right click (control click for Mac) on Saturn and select **go there**. You can animate the journey by first unchecking the **only animate intra planet changes** box under **file/preferences/ responsiveness** (also, be sure to have your horizon turned on to see the photorealistic surface panorama). Click the **W viewing direction** button on the **button bar**, or simply hit the W key on the keyboard. Select a **time speed** of **1 minute**. Click the PLAY **time mode** button, and click STOP when the Sun is near setting.** Use the FORWARD STEP and BACKWARD STEP **time mode** buttons until you see the Sun just starting to set. You may then need to select a smaller unit of time until the Sun is just touching the horizon (Saturn has no solid surface, so it's really the cloud tops we are viewing). Note the date and time, then continue on to the next sunset and calculate the time difference between them.

Date and time of first sunset on Saturn:	
Date and time of second sunset on Saturn:	
Subtract to get length of solar day:	

*For even greater accuracy, instead of using sunset, you should use solar noon. To find solar noon, use the LOCAL (lo) **coordinate system** button to align the Sun with the local meridian line, the Sun's highest point in the sky for that day. Starry Night™ Pro also gives the time of solar noon in the **date & time** drop-down menu on the **control panel**.

If you are unable to see the Sun to the west during sunset, first use the **date & time drop-down menu to select **sunset**, then click the check box to the left of the Sun in the FIND **side pane** to label the Sun, then pan to the left (SW) or right (NW) until you see it. If you still have trouble locating the Sun, right click (control click for Mac) on the Sun in the FIND **side pane** and select **centre**. You will need to unlock the Sun before continuing with this activity. The easiest way to "unlock" is to simply "grab" the sky by clicking and holding the left mouse button, and then moving the mouse a little bit in any direction.

The solar day is not the same as the sidereal rotation period of a planet. This is because the solar day takes into account both the rotation of the planet and the revolution of the planet around the Sun. Your calculation of a solar day is, therefore, a little different than the sidereal day found earlier in this activity from the INFO **side pane**.

How do we determine the revolutionary period, or the planet's "year"?

Go to the **solar system** view. From the **menu**, select **favourites/solar system/outer planets/** and select **outer solar system**. Click the STOP **time mode** button, then select **now** from the **date and time** drop-down menu. Choose a **time speed** of **1 day**, click the PLAY **time mode** button, then note the time it takes for the planet to return to its original position (you may need to increase the number of days depending on the speed of your computer). You may find it easiest to align the planet to the farthest left or right before starting. Complete the chart below.

Date of Saturn at start:	
Date of Saturn after one revolution:	
Subtract to get length of year:	

Activity 12—SATURN'S MOONS

*These activities are designed to work with the Starry Night software that comes with your text, from any home location you choose, and with the current date and time, unless indicated otherwise. You may always revert to factory default settings by clicking **FILE/ preferences**, then selecting **factory defaults** as needed. You may also undo a command or series of commands on the PC by clicking the **back** button at the top left of the **button bar**. You should refer to the key given at the beginning of this booklet for clarification of "on screen" buttons, controls, and functions. PC **button bar** items can all be accessed through the **menu**. "Right click" on the PC is equivalent to "control click" on the Mac. All activities assume that OpenGL graphics capabilities are enabled on your computer.*

PART 1: FINDING SATURN

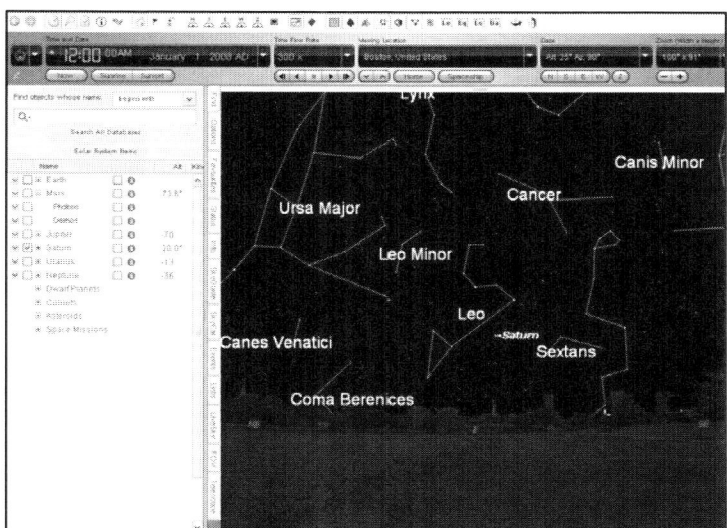

You should begin this activity at sunset. An easy way to do this is to click the drop-down menu to the right of the **date & time** field on the **control panel** and select **sunset**. Look toward the west by clicking the **W viewing direction** button located on the **button bar** across the top of your screen, or by simply keying in the letter W (Mac users should refer to the button bar commands given at the beginning of this booklet). The screen will pan toward the west. Select a playing speed of **300×** normal time by clicking the drop-down menu at the right of the **time speed** field. Click the STOP **time mode** button when the Sun has set, the stars have come out, and dusk is almost over. Then click on the **constellations** button to show the constellations.

Click the FIND tab on the left **side pane**. A list of planets should appear. Those that are highlighted are currently up in your evening sky. Those that are not highlighted are not up in the sky at this time. We wish to find the planet Saturn. If Saturn is highlighted (see footnote if not highlighted)*, double-click on it or you may right click (control click for Mac) and select **center**. This will pan the screen and center Saturn. You can now zoom in on Saturn either by using the **zoom control** at the far right of the **control panel**, or by right clicking (control clicking for Mac) on the highlighted listing in the FIND **side pane** and selecting **magnify**.

*If Saturn is not currently highlighted, you will need to move time forward to a time when Saturn will rise in the east. Start by looking toward the east by clicking the **E viewing direction** button located on the **button bar** across the top of your screen. Select a playing speed of **300×** or **3000×** normal time by clicking the drop-down menu at the right of the **time speed** field. Click the STOP **time mode** button when the planet Saturn becomes highlighted in the FIND **side pane** listing. It may be that Saturn is up during the day. If the Sun rises before Saturn does, then click the **daylight** button on the **button bar**. This will keep the sky dark so you can see the stars and constellations. Once Saturn has risen, double-click on it. This will pan the screen and center Saturn. You can now zoom in on Saturn either by using the **zoom control** at the far right of the **control panel**, or by right clicking (control clicking for Mac) on the highlighted listing in the FIND **side pane** and selecting **magnify**.

Click the information icon **(i)** in the **side pane** to read a short description of the planet Saturn. For even more information, select the INFO tab on the left **side pane** and click the plus sign (gray arrow for Mac) to expand the different information categories.

PART 2: SATURN'S MOONS TITAN AND MIMAS

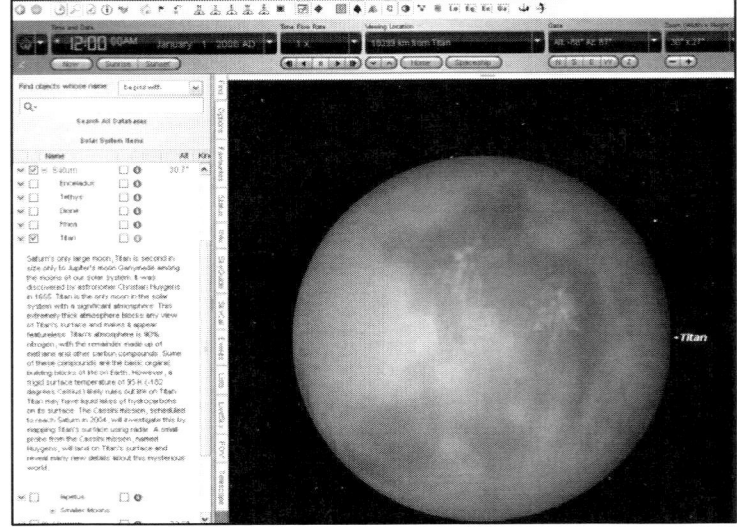

We wish to take a closer look at Saturn's largest moon—Titan, and a small but interesting moon—Mimas. Titan has a methane atmosphere that obscures the surface. The organic molecules on this moon may indicate conditions suitable for life. Mimas is interesting because of a very large impact crater a quarter of the diameter of the moon itself. This is one of the largest crater-to-moon size ratios in the solar system and could have shattered Mimas into pieces.

Click the plus sign (gray arrow for Mac) next to **Saturn** in the FIND **side pane**. This will list Saturn's moons. Click the first check box for both **Titan** and **Mimas** to label them on your screen. In order to see both the planet and the moons, you may need to pan out by using the **zoom control** at the far right of the **control panel**. Keep in mind that the moons may be out of sight behind the planet itself. In that case, use the **3000× time speed** setting on the **control panel** to move time forward so that both moons are clearly visible. Remember that as you move time forward, you may need to click the **horizon** and **daylight** buttons on the **button bar** as needed.

Let's take a closer look at Titan and Mimas. Double-click on **Titan** in the list. This will center Titan on your screen. Use the **zoom control** to get a close up view.

[OPTIONAL] If you would like to get an even closer view beyond the limits of the zoom function, right click (control click for Mac) on the moon and select **go there**. You are now on the surface (to see the surface, make sure your horizon is turned on). Now launch yourself up above the moon by clicking the **location above surface of planet** up arrow on the **control panel**. Lock onto the moon by double-clicking on it in the FIND **side pane**, then use the **zoom control** as needed to see the moon full screen. When done, be sure to click the **back** button on the **button bar** a few times to return to an earthbound perspective for the rest of the assignment.

Click the information icon **(i)** in the **side pane** to read a short description of this moon. Then, select the INFO tab on the left **side pane** and click the plus sign (gray arrow for Mac) to expand the different information categories. Do the same to view Mimas. See if you can see the exceptionally large impact crater Mimas is so famous for.

Complete the following table:

Compare to Earth's Moon with radius = 1737 km, and mass = 7.35×10^{22} kg = 0.0123 Earth masses. 1 AU = 1.50×10^{8} km. Saturn's radius is 60267 km, and Saturn's mass is 95.16 Earth masses. Earth's mass = 5.974×10^{24} kg.

ATTRIBUTE	TITAN	MIMAS
Radius (in km):		
…in multiples of Earth's Moon radius*:		
Mass (in Earth masses):		
…in multiples of Earth's Moon mass:		
…as ratio of Saturn's mass:		
Orbit size (in AU):		
…in km:		
…in multiples of Saturn's radius:		
Sidereal day (one rotation):		
Solar day (noon to noon):		
Year (to orbit around Saturn):		

*The info given for *radius* is in km. To compute the *multiple of Earth's Moon radius*, take *radius* in km and divide by *Earth's Moon radius* as given above. Follow the same procedure for *mass* and *orbit size*.

Note how small the ratio of Titan's mass is as compared to the mass of Saturn itself. Although Titan is the second largest moon in the solar system, its mass ratio is very small, with little gravitational effect on Saturn. In contrast, Earth's moon-to-planet mass ratio is one of the largest in the solar system at 0.0123 (or 1.23%).

Note how Saturn's rings do not extend beyond a few multiples of Saturn's radius. Compare this with our own Moon, whose distance from Earth is over 60 times Earth's radius. When a large moon is too close to its parent planet, the tidal forces are so great that they try to rip the moon apart. The critical distance, known as the Roche limit, is the closest distance a large moon can exist in orbit around a planet without being ripped to shreds. Moons are thus generally found outside the Roche limit, and ring systems are usually found within the Roche limit. The origin of Saturn's rings is still under debate. The particles from which they are made may either have tried to come together over the ages to form a moon, but been unable to do so because they are within the Roche limit; a large moon, meteoroid, or comet may have entered within the Roche limit and been ripped to shreds, scattering the debris that currently comprises today's ring system. Statistically, Saturn's original ring system should have dissipated by now, so the question is how are the rings being maintained or rejuvenated?

PART 3: SATELLITE ORBITAL CHARACTERISTICS

Zoom back out so you can see Saturn and its moons once again. Click the **labels** button on the **button bar** to see the names of other moons. Double-click on **Saturn** in the FIND **side pane** to center on Saturn, and zoom in and out a bit to see how many moons you can observe. Select the **3000× time speed** on the **control panel** and watch the moons orbit Saturn. Click the **daylight** and **horizon** buttons as needed. For best results, so that the smaller moons will be displayed, Saturn should cover about a third of your screen. Watch the motion for a bit. The tilting is due to our local perspective. You can remove this effect by selecting **view/ecliptic guides**, then **the ecliptic**. Why do the moons move from side-to-side rather than in circles around Saturn? This also has to do with perspective. Since Saturn's satellite orbital plane lies along the ecliptic, the plane of our solar system, we can only observe Saturn from the side.

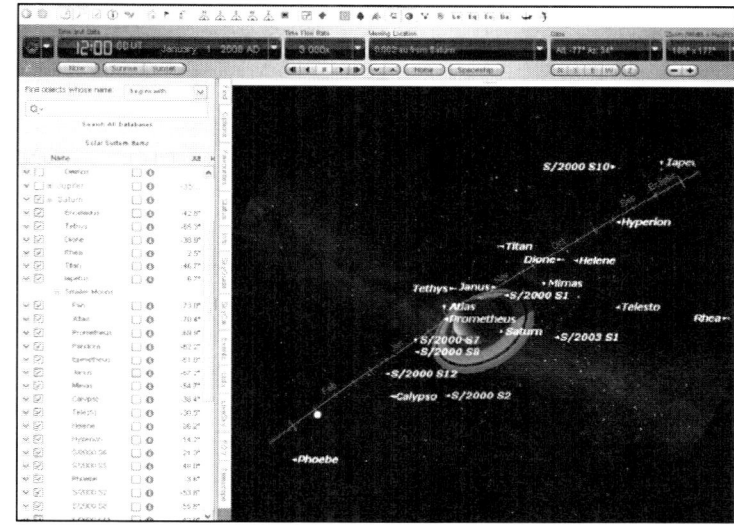

[OPTIONAL] You can use Starry Night™ Pro to observe the moons from above Saturn's polar region. First, to go to Saturn, then change your position to Saturn's north pole. Next, launch yourself straight up until you can view the moons from above. Now, when you click the **play** button, you will see the moons orbiting in nearly circular orbits. When done, click the **back** button repeatedly until you have returned to your original perspective from Earth.

Let's take a closer look at satellite orbits. To better visualize the satellite orbital motion, click both the first and second set of check boxes in the FIND **side pane** for **Titan** and a few other moons of your choosing (for example, **Mimas**) to label and trace their respective orbits around Saturn. Again, you may need to adjust the **zoom controls** to better view the orbital plane. To get a feel for how fast these moons move around in their orbit, slowly move time forward using the **300×** or **3000× time speeds** and observe the orbital motion as it would appear as seen from Earth.

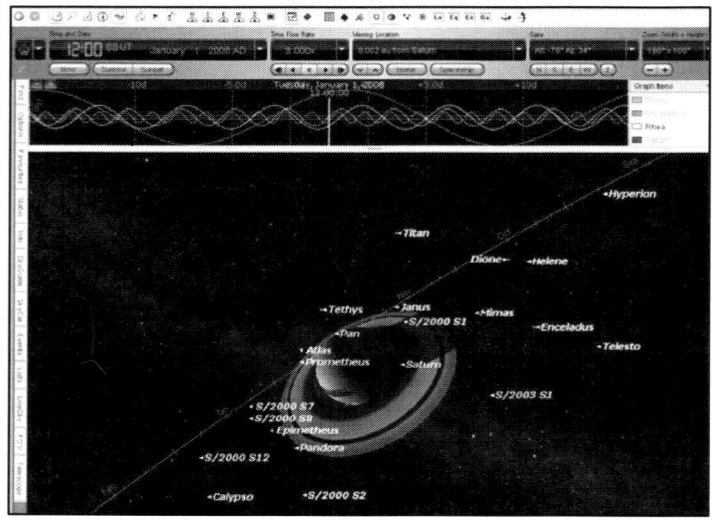

To see some different viewing perspectives of the moons, right click (control click for Mac) on Saturn and select **graph elongation of moons**. The graph shows the angular separation between numerous moons and Saturn. Titan is graphed, but Mimas is not. To add Mimas to the graph, right click (control click for Mac) on Mimas either in the FIND **side pane** or on the screen and select **start graphing**. Similarly, you can remove moons from the graph by right clicking (control clicking for Mac) on them and selecting **stop graphing**. You can move time forward with the **time speed** controls on the **control panel**, or you can simply grab the graph by clicking down on it with the mouse and dragging to the left or right while holding the mouse button down. You can also expand the horizontal resolution of the graph by clicking on the plus and minus indicators at the top left of the graph. To increase the vertical resolution, grab the bottom of the graph with the mouse and pull down. By moving the graph back and forth to look for different alignments, you can answer the following questions.

SATELLITE ORBITAL CHARACTERISTICS:

Time for a complete cycle of Titan:		
Maximum elongation for Titan (see vertical scale)*:		
Time for a complete cycle of Mimas:		
Maximum elongation for Mimas (see vertical scale)*:		

*You can check and improve on the precision of the maximum elongation values by centering your pointer on Saturn, then clicking down and dragging the pointer to the farthest extent of the moon's orbit (make sure the orbit trace is on). The angular separation will be displayed.

Activity 13—THE PLANET URANUS

*These activities are designed to work with the Starry Night software that comes with your text, from any home location you choose, and with the current date and time, unless indicated otherwise. You may always revert to factory default settings by clicking **FILE/ preferences**, then selecting **factory defaults** as needed. You may also undo a command or series of commands on the PC by clicking the **back** button at the top left of the **button bar**. You should refer to the key given at the beginning of this booklet for clarification of "on screen" buttons, controls, and functions. PC **button bar** items can all be accessed through the **menu**. "Right click" on the PC is equivalent to "control click" on the Mac. All activities assume that OpenGL graphics capabilities are enabled on your computer.*

PART 1: FINDING URANUS

You should begin this activity at sunset. An easy way to do this is to click the drop-down menu to the right of the **date & time** field on the **control panel**, and select **sunset**. Look toward the west by clicking the **W viewing direction** button located on the **button bar** across the top of your screen, or by simply keying in the letter W (Mac users should refer to the button bar commands given at the beginning of this booklet). The screen will pan toward the west. Select a playing speed of **300×** normal time by clicking the drop-down menu at the right of the **time speed** field. Click the STOP **time mode** button when the Sun has set, the stars have come out, and dusk is almost over. Then click on the **constellations** button to show the constellations.

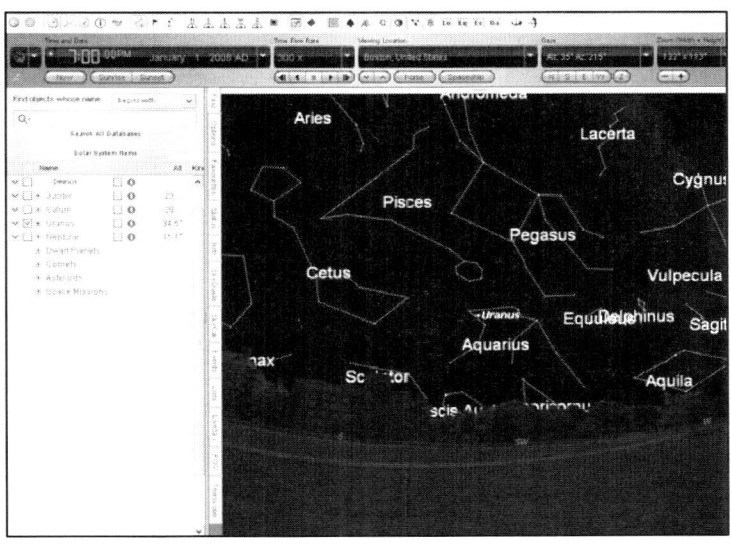

Click the FIND tab on the left **side pane**. A list of planets should appear. Those that are highlighted are currently up in your evening sky. Those that are not highlighted are not up in the sky at this time. We wish to find the planet Uranus. If Uranus is highlighted (see footnote if not highlighted)*, double click or right click on it (control click for Mac) and select **center**. This will pan the screen and center Uranus. You can now zoom in on Uranus either by using the **zoom control** at the far right of the **control panel**, or by right clicking (control clicking for Mac) on the highlighted listing in the FIND **side pane** and selecting **magnify**.

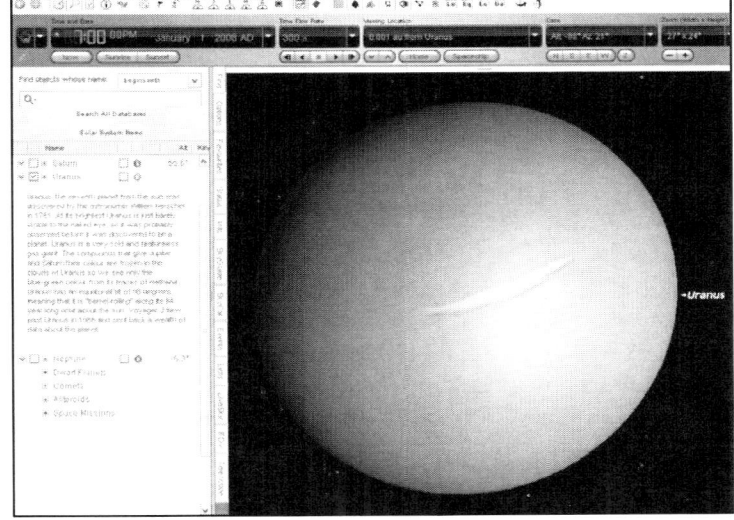

*If Uranus is not currently highlighted, you will need to move time forward to a time when Uranus will rise in the east. Start by looking toward the east by clicking the **E viewing direction** button located on the **button bar** across the top of your screen. Select a playing speed of 300× or 3000× normal time by clicking the drop-down menu at the right of the **time speed** field. Click the STOP **time mode** button when the planet Uranus becomes highlighted in the FIND **side pane** listing. It may be that Uranus is up during the day. If the Sun rises before Uranus does, then click the **daylight** button on the **button bar**. This will keep the sky dark so you can see the stars and constellations. Once Uranus has risen, double-click on it. This will pan the screen and center Uranus. You can now zoom in on Uranus either by using the **zoom control** at the far right of the **control panel**, or by right clicking (control clicking for Mac) on the highlighted listing in the FIND **side pane** and selecting **magnify**.

Click the information icon **(i)** in the **side pane** to read a short description of the planet Uranus. Next, select the INFO tab on the left **side pane** and click the plus sign (gray arrow for Mac) to expand the different information categories.

Use the information in the OTHER DATA section to complete the following table of Uranus's physical characteristics and compare to Earth.
(You will need to know Earth's radius = 6378 km. To compute percent, divide one value by the other and multiply by 100.)

PHYSICAL CHARACTERISTICS:

Radius of Uranus:	_____ km, _____ % of Earth
Mass of Uranus:	
Angular size in arc minutes and seconds as seen from Earth:	
Orbit size (mean distance from the Sun):	
Sidereal day (period of rotation):	
Solar day (noon to noon on surface):	
Length of year:	

Complete the following table of Uranus's observational characteristics:*

OBSERVATIONAL CHARACTERISTICS:

Date, time:	
Azimuth, altitude:	
Currently in which constellation:	
Current apparent magnitude:	
Max possible magnitude as seen from Earth:	
Current distance from Earth:	
Distance at conjunction (aligned with the sun):	
Distance at opposition (opposite the sun):	

*For *constellation, azimuth, and altitude,* see POSITION IN SKY; for *Distance from Earth,* see POSITION IN SPACE. For everything else, see OTHER DATA. To compute *distance at conjunction,* take *orbit size* and add Earth's distance from the Sun of 1 AU. To compute *distance at opposition,* take *orbit size* and subtract Earth's distance from Sun of 1 AU.

PART 2: URANUS'S ORBITAL CHARACTERISTICS

Use the **zoom control** at the far right of the **control panel** to zoom back out to full-scale view. Right click (control click for Mac) on Uranus and select **orbit**. This shows Uranus's orbital path as seen from Earth. You will need to click the **horizon** button on the **button bar** to get rid of the Earth's horizon should it interfere with your view of Uranus's orbit. You should also click the **daylight** button to turn off daylight. You should still be locked on to Uranus. If not, right click (control click for Mac) on the planet Uranus, and then select **center**. To maintain the proper perspective, select **view/ecliptic guides**, then **the ecliptic**.

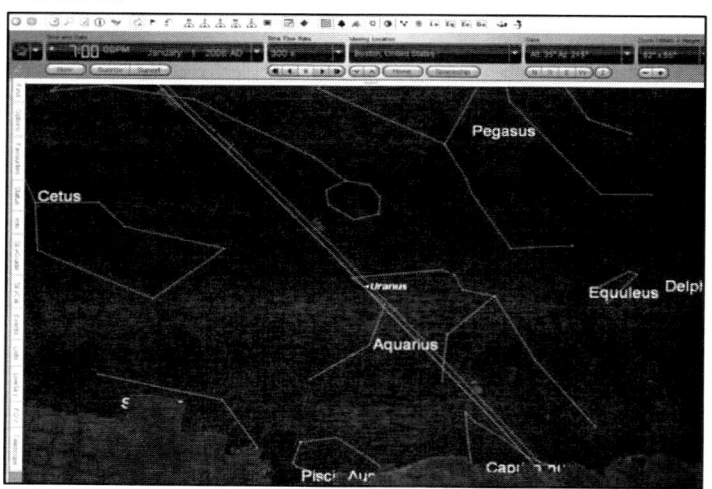

Start with today's date by entering it in the **date & time** field of the **control panel**, or by selecting **now** from the drop-down menu to the right of the **date & time** field. Note what constellation Uranus is in, then select a **time speed** of **1 day** and click the PLAY **time mode** button. If you need to slow down or speed up, adjust the **time speed** fields as needed. You should be able to see Uranus locked in the center of your field of view, but moving across the star background. You should also be able to complete a full revolution in less than a minute. Click the STOP **time mode** button when Uranus has returned to the same constellation and note the date. Use this information to complete the following table.

VIEWING URANUS'S RELATIVE POSITION:

Constellation Uranus is located in today:	
Today's date (corresponding to location of Uranus given above):	
Date when Uranus is again located in same constellation:	
Length of time it takes Uranus to return to the same place in the sky (subtract the dates given above):	

*Note that since Uranus is relatively far from the Sun as compared to the Earth, this value will be close to Uranus's sidereal period (year). However, there will be a difference since we are looking at Uranus's position relative to the Earth, which is itself in an orbit around the Sun.

PART 3: FLYING TO URANUS

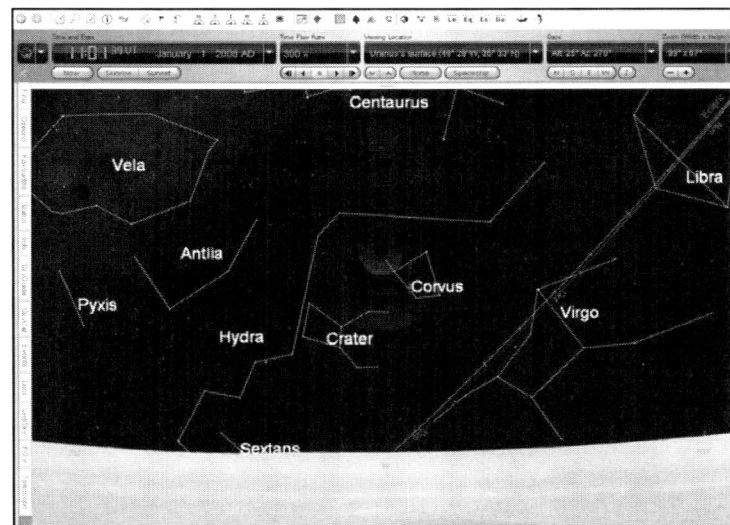

Let's see what a day would be like on Uranus. Right click (control click for Mac) on Uranus and select **go there**. You can animate the journey by first unchecking the **only animate intra planet changes** box under **file/preferences/responsiveness** (also, be sure to have your horizon turned on to see the photorealistic surface panorama). Click the **W viewing direction** button on the **button bar**, or simply hit the W key on the keyboard. Select **3000× time speed**. Click the PLAY **time mode** button, and observe what a day would look like as seen from the surface of Uranus (actually, Uranus has no solid surface, so it's really the cloud tops we are viewing). Note that due to Uranus's retrograde rotation, the motion appears to be backwards when compared to what we are used to here on Earth (and as it appears for most of the other planets).

[OPTIONAL] Does the Sun rise and set as it does on Earth? Depending on where Uranus is in its orbit, the Sun may rise and set in a daily cycle like on Earth (near to Uranus's equinox dates), remain high in the sky for decades (near to Uranus's summer solstice), or be completely out of view for decades (near to Uranus's winter solstice). As an optional activity, you may input different dates in the **date & time** field and try and determine when and why this occurs. Be sure to take note of the length of Uranus's year as determined below, or as given in the INFO **side pane**.

How do we determine the revolutionary period, or the length of a planet's "year"? Go to the solar system view. From the **menu**, select **favourites/solar system/outer planets/** and select **outer solar system**. Click the STOP **time mode** button, then select **now** from the **date and time** drop-down menu. Choose a **time speed** of **1 day**, click the PLAY **time mode** button, then note the time it takes for the planet to return to its original position (you may need to increase the number of days depending on the speed of your computer). You may find it easiest to align the planet to the farthest left or right before starting. Complete the chart below.

Date of Uranus at start:	
Date of Uranus after one revolution:	
Subtract to get length of year:	

Activity 14—THE MOONS of URANUS

*These activities are designed to work with the Starry Night software that comes with your text, from any home location you choose, and with the current date and time, unless indicated otherwise. You may always revert to factory default settings by clicking **FILE/ preferences**, then selecting **factory defaults** as needed. You may also undo a command or series of commands on the PC by clicking the **back** button at the top left of the **button bar**. You should refer to the key given at the beginning of this booklet for clarification of "on screen" buttons, controls, and functions. PC **button bar** items can all be accessed through the **menu**. "Right click" on the PC is equivalent to "control click" on the Mac. All activities assume that OpenGL graphics capabilities are enabled on your computer.*

PART 1: FINDING URANUS

You should begin this activity at sunset. An easy way to do this is to click the drop-down menu to the right of the **date & time** field on the **control panel**, and select **sunset**. Look toward the west by clicking the **W viewing direction** button located on the **button bar** across the top of your screen, or by simply keying in the letter W (Mac users should refer to the button bar commands given at the beginning of this booklet). The screen will pan toward the west. Select a playing speed of **300×** normal time by clicking the drop-down menu at the right of the **time speed** field. Click the STOP **time mode** button when the Sun has set, the stars have come out, and dusk is almost over. Then click on the **constellations** button to show the constellations.

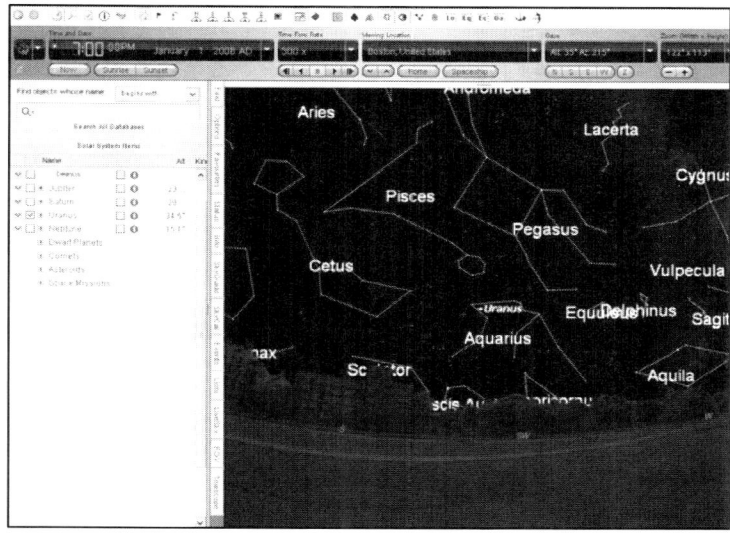

Click the FIND tab on the left **side pane**. A list of planets should appear. Those that are highlighted are currently up in your evening sky. Those that are not highlighted are not up in the sky at this time. We wish to find the planet Uranus. If Uranus is highlighted (see footnote if not highlighted)*, double-click on it or you may right click (control click for Mac) and select **centre**. This will pan the screen and center Uranus. You can now zoom in on Uranus either by using the **zoom control** at the far right of the **control panel**, or by right clicking (control clicking for Mac) on the highlighted listing in the FIND **side pane** and selecting **magnify**.

*If Uranus is not currently highlighted, you will need to move time forward to a time when Uranus will rise in the east. Start by looking toward the east by clicking the **E viewing direction** button located on the **button bar** across the top of your screen. Select a playing speed of **300×** or **3000×** normal time by clicking the drop-down menu at the right of the **time speed** field. Click the STOP **time mode** button when the planet Uranus becomes highlighted in the FIND **side pane** listing. It may be that Uranus is up during the day. If the Sun rises before Uranus does, then click the **daylight** button on the **button bar**. This will keep the sky dark so you can see the stars and constellations. Once Uranus has risen, double-click on it. This will pan the screen and center Uranus. You can now zoom in on Uranus either by using the **zoom control** at the far right of the **control panel**, or by right clicking (control clicking for Mac) on the highlighted listing in the FIND **side pane** and selecting **magnify**.

Click the information icon **(i)** in the **side pane** to read a short description of the planet Uranus. For even more information, select the INFO tab on the left **side pane** and click the plus sign (gray arrow for Mac) to expand the different information categories.

PART 2: URANUS'S MOON MIRANDA

We wish to take a closer look at Uranus's moon Miranda.

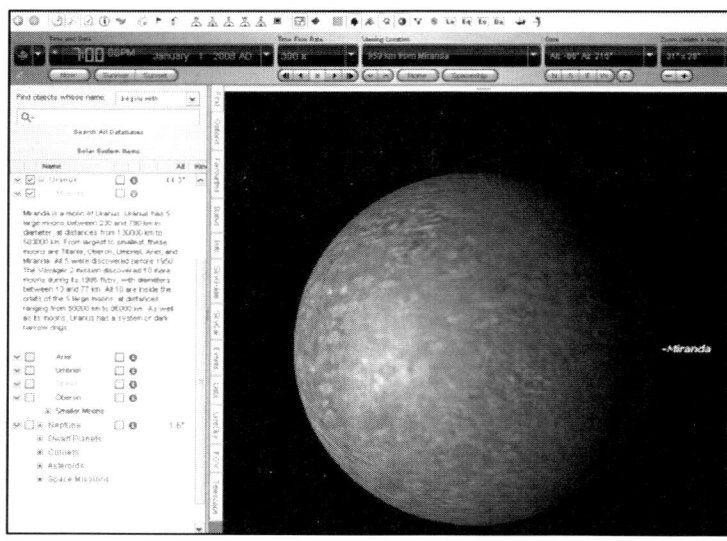

Click the plus sign (gray arrow for Mac) next to **Uranus** in the FIND **side pane**. This will list Uranus's moons. Click the first check box for **Miranda** to label it on your screen. In order to see both the planet and the moon, you may need to pan out by using the **zoom control** at the far right of the **control panel**. Keep in mind that Miranda may be out of sight behind the planet itself. In that case, use the **3000× time speed** setting on the **control panel** to move time forward so that Miranda is clearly visible. Remember that as you move time forward, you may need to click the **horizon** and **daylight** buttons on the **button bar** as needed.

Let's take a closer look at Miranda. Double click on **Miranda** in the list. This will center Miranda on your screen. Use the **zoom control** to get a close up view.

[OPTIONAL] If you would like to get an even closer view beyond the limits of the zoom function, right click (control click for Mac) on the moon and select **go there**. You are now on the surface (to see the surface, make sure your horizon is turned on). Now launch yourself up above the moon by clicking the **location above surface of planet "up"** arrow on the **control panel**. Lock onto Miranda by double-clicking on **Miranda** in the FIND **side pane**, then use the **zoom control** as needed to see the moon full screen. When done, be sure to click the **back** button on the **button bar** a few times to return to an earthbound perspective for the rest of the assignment.

Click the information icon **(i)** in the **side pane** to read a short description of this moon. Then, select the INFO tab on the left **side pane** and click the plus sign (gray arrow for Mac) to expand the different information categories.

Complete the following table:

Compare to Earth's Moon with radius = 1737 km, and mass = 7.35×10^{22} kg = 0.0123 Earth masses.

Uranus's radius is 25558 km, Uranus's mass is 14.544 Earth masses, and Earth's mass = 5.974×10^{24} kg.

ATTRIBUTE	MIRANDA
Radius (in km):	
…in multiples of Earth's Moon radius:*	
Mass (in Earth masses):	
…in multiples of Earth's Moon mass:	
…as ratio of Uranus's mass:	
Orbit size (in km):	
…in multiples of Uranus's radius:	
Sidereal day (one rotation):	
Solar day (noon to noon):	
Year (to orbit around Uranus):	

*The info given for *radius* is in km. To compute the *multiple of Earth's Moon radius* take *radius* in km and divide by *Earth's Moon radius* as given above. Follow the same procedure for *mass* and *orbit size*.

Note how small the ratio of Miranda's mass is as compared to the mass of Uranus itself. Its mass ratio is very small, with little gravitational effect on Uranus. In contrast, Earth's moon-to-planet-mass ratio is one of the largest in the solar system at 0.0123 (or 1.23%).

PART 3: SATELLITE ORBITAL CHARACTERISTICS

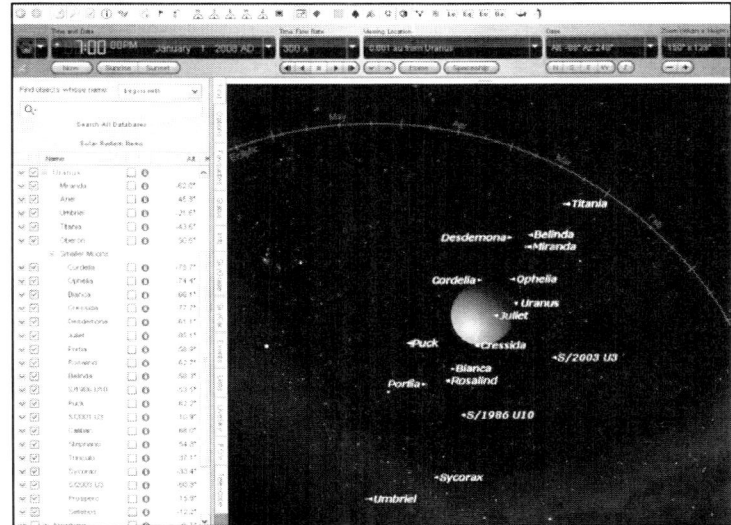

Zoom back out so you can see Uranus and its moons once again. Click the **labels** button on the **button bar** to see the names of other moons. Double click on **Uranus** in the FIND **side pane** to center on Uranus and zoom in and out a bit to see how many moons you can observe. Select the **3000× time speed** on the **control panel** and watch the moons orbit Uranus. Click the **daylight** and **horizon** buttons as needed. For best results, so that the smaller moons will be displayed, Uranus should cover about a third of your screen. Watch the motion for a bit. The tilting is due to our local perspective. You can remove this effect by selecting **view/ecliptic guides**, then **the ecliptic**.

Let's take a closer look at satellite orbits. To better visualize the satellite orbital motion, click both the first and second set of check boxes in the FIND **side pane** for **Miranda** and a few other moons of your choosing (for example, **Portia**) to label and trace their respective orbits around Uranus. Again, you may need to adjust the **zoom controls** to better view the orbital plane. To get a feel for how fast these moons move around in their orbit, slowly move time forward using the **300×** or **3000× time speed** and observe the orbital motion as it would appear as seen from Earth.

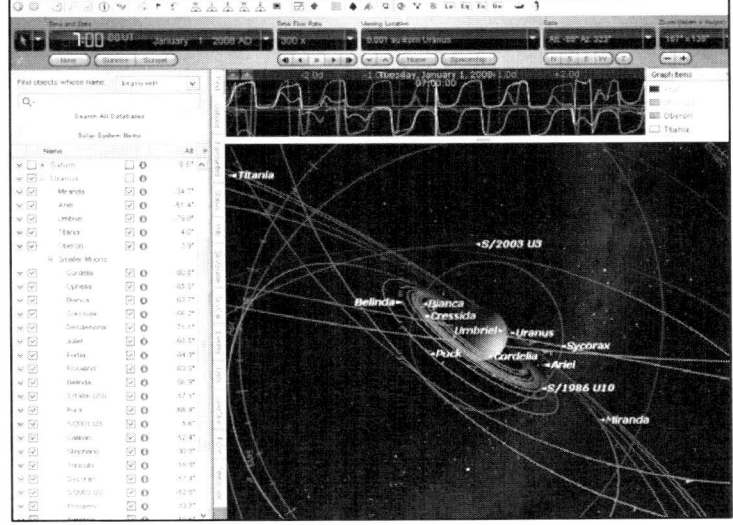

To see some different viewing perspectives of the moons, right click (control click for Mac) on Uranus and select **graph elongation of moons**. The graph shows the angular separation between numerous moons with Uranus. Miranda is graphed, but Portia is not. To add Portia to the graph, right click (control click for Mac) on Portia either in the FIND **side pane** or on the screen and select **start graphing**. Similarly, you can remove moons from the graph by right clicking (control clicking for Mac) on them and selecting **stop graphing**. You can move time forward with the **time speed** controls on the **control panel**, or you can simply grab the graph by clicking down on it with the mouse, then dragging to the left or right while holding the mouse button down. You can also expand the horizontal resolution of the graph by clicking on the plus and minus indicators at the top left of the graph. To increase the vertical resolution, grab the bottom of the graph with the mouse and pull down. By moving the graph back and forth to look for different alignments, you can answer the following questions.

SATELLITE ORBITAL CHARACTERISTICS:

Time for a complete cycle of Miranda	
Maximum elongation for Miranda (see vertical scale)*	
Time for a complete cycle of Portia	
Maximum elongation for Portia (see vertical scale)*	

*You can check and improve on the precision of the maximum elongation values by centering your pointer on Uranus (make sure the orbit trace is on), then clicking down and dragging the pointer to the farthest extent of the moon's orbit (make sure the orbit trace is on). The angular separation will be displayed.

PART 4: A PLANET TILTED ON ITS SIDE

You should notice that unlike the other planets, the satellite orbits do not align with our ecliptic (depending on the time of year). To see this, select **view/ecliptic guides**, then **the ecliptic**. Also, click the ECLIPTIC **coordinate system** button on the **button bar** (for Macs, refer to the button bar commands given at the beginning of this booklet). The horizontal grid lines are now aligned with the Earth's orbit around the Sun. Although many of the planets are slightly tilted with respect to the ecliptic, Uranus's tilt is extreme. The planet (and corresponding satellite orbits) are tilted almost completely on their side. This results in a unique observational effect in that, at certain times during Uranus's year, at certain positions in its orbit, the satellite orbits form a bull's-eye as seen from Earth.

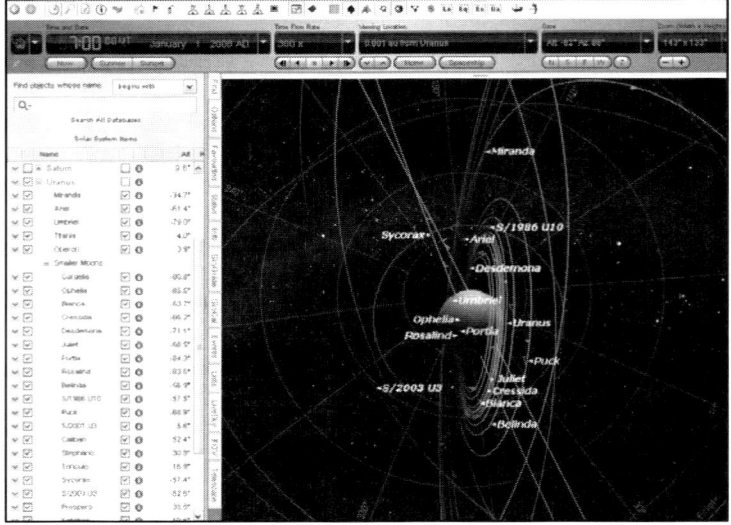

To observe this phenomenon, lock on to Uranus by double-clicking **Uranus** in the FIND **side pane**. Check both boxes for all the moons. Zoom in and out so you can see all the satellite orbits. First note that the outer moons have more eccentric orbits and do not align with Uranus's rotational plane. The inner moons, however, align nicely with the rotational plane. Zoom in so that you can best view the inner moons that are aligned. On the **control panel**, select a **time speed** of **1 year**, and click the PLAY **time mode** button. You should see an animation of the moons orbiting Uranus while Uranus changes its perspective relative to us on Earth. You should note how, at certain times of the year, you get a bull's-eye pattern.

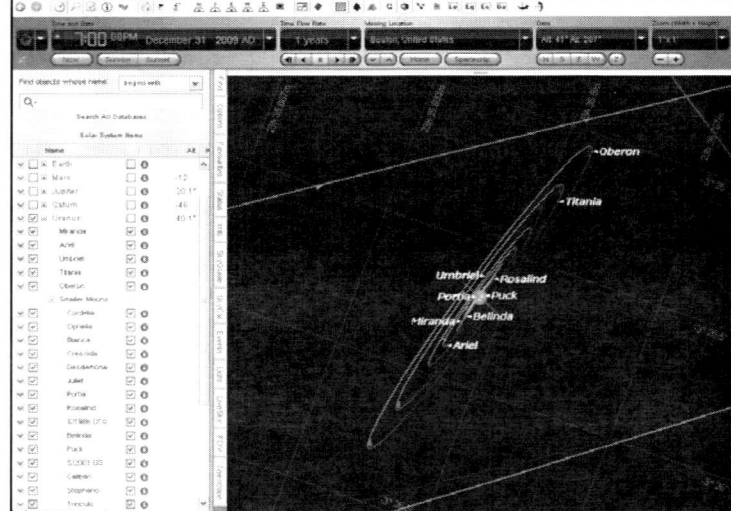

The animation may seem a bit bouncy. Each "bounce" is approximately an Earth year, bringing us a little closer, and then taking us away again from Uranus. Uranus is 20 AU away, and the Earth's orbit changes the distance by 2 AU every six months. It's only a 10% difference, but notable enough to cause the "bounces."

[OPTIONAL] You can remove the "bounce" effect by using Starry Night™ Pro to observe from the Sun rather than from Earth. First go there, and then lock onto Uranus from this new position. Now when you click the **play** button, you can observe Uranus's orbital motion, and it will appear smooth (without the bounce due to Earth's orbital motion). When done, click the **back** button repeatedly until you have returned to your original perspective from Earth.

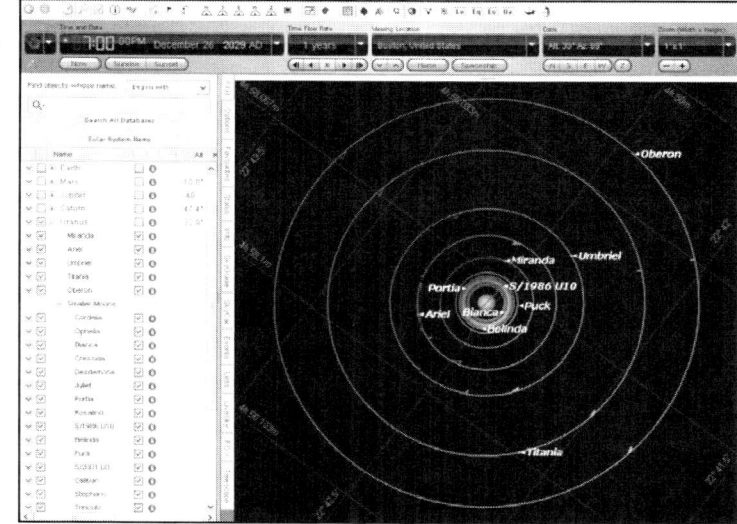

Once you have experimented a bit, answer the following questions:

EVENT	DATE
When will Uranus next form a bull's-eye:	
When will Uranus next be viewed edge on:	
The following time Uranus will form a bull's-eye:	
The following time Uranus will be viewed edge on:	

Activity 15—THE PLANET NEPTUNE

*These activities are designed to work with the Starry Night software that comes with your text, from any home location you choose, and with the current date and time, unless indicated otherwise. You may always revert to factory default settings by clicking **FILE/ preferences**, then selecting **factory defaults** as needed. You may also undo a command or series of commands on the PC by clicking the **back** button at the top left of the **button bar**. You should refer to the key given at the beginning of this booklet for clarification of "on screen" buttons, controls, and functions. PC **button bar** items can all be accessed through the **menu**. "Right click" on the PC is equivalent to "control click" on the Mac. All activities assume that OpenGL graphics capabilities are enabled on your computer.*

PART 1: FINDING NEPTUNE

You should begin this activity at sunset. An easy way to do this is to click the drop-down menu to the right of the **date & time** field on the **control panel**, and select **sunset**. Look toward the west by clicking the **W viewing direction** button located on the **button bar** across the top of your screen, or by simply keying in the letter W (Mac users should refer to the button bar commands given at the beginning of this booklet). The screen will pan toward the west. Select a playing speed of **300×** normal time by clicking the drop-down menu at the right of the **time speed** field. Click the STOP **time mode** button when the Sun has set, the stars have come out, and dusk is almost over. Then click on the **constellations** button to show the constellations.

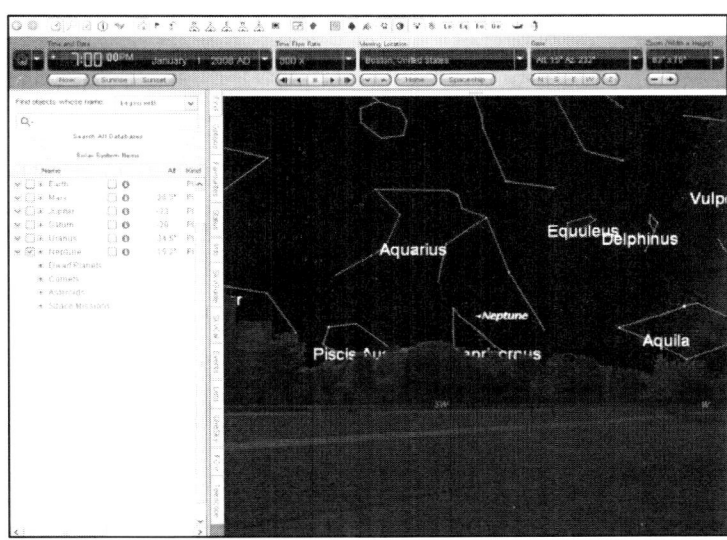

Click the FIND tab on the left **side pane**. A list of planets should appear. Those that are highlighted are currently up in your evening sky. Those that are not highlighted are not up in the sky at this time. We wish to find the planet Neptune. If Neptune is highlighted (see footnote if not highlighted)*, double-click or right click on it (control click for Mac) and select **centre**. This will pan the screen and center Neptune. You can now zoom in on Neptune either by using the **zoom control** at the far right of the **control panel**, or by right clicking (control clicking for Mac) on the highlighted listing in the FIND **side pane** and selecting **magnify**.

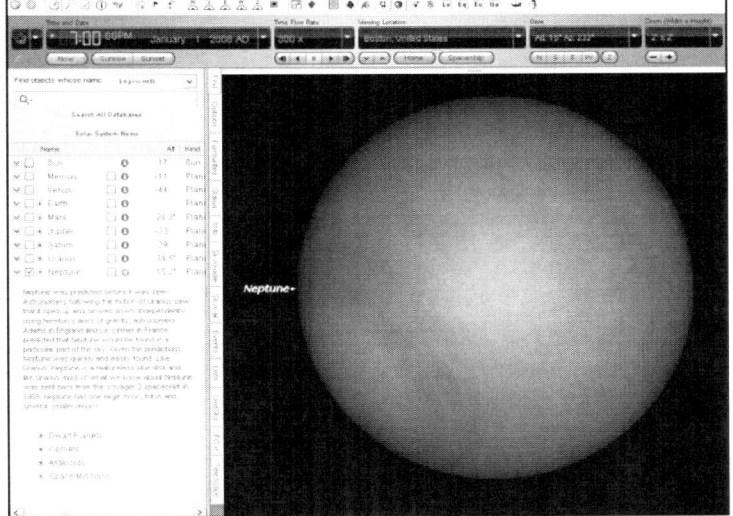

*If Neptune is not currently highlighted, you will need to move time forward to a time when Neptune will rise in the east. Start by looking toward the east by clicking the **E viewing direction** button located on the **button bar** across the top of your screen. Select a playing speed of **300×** or **3000×** normal time by clicking the drop-down menu at the right of the **time speed** field. Click the STOP **time mode** button when the planet Neptune becomes highlighted in the FIND **side pane** listing. It may be that Neptune is up during the day. If the Sun rises before Neptune does, then click the **daylight** button on the **button bar**. This will keep the sky dark so you can see the stars and constellations. Once Neptune has risen, double-click on it. This will pan the screen and center Neptune. You can now zoom in on Neptune either by using the **zoom control** at the far right of the **control panel**, or by right clicking (control clicking for Mac) on the highlighted listing in the FIND **side pane** and selecting **magnify**.

Click the information icon **(i)** in the **side pane** to read a short description of the planet Neptune. Next, select the INFO tab on the left **side pane** and click the plus sign (gray arrow for Mac) to expand the different information categories.

Use the information in the OTHER DATA section to complete the following table of Neptune's physical characteristics and compare to Earth:

(You will need to know Earth's radius = 6378 km. To compute percent, divide one value by the other and multiply by 100.)

PHYSICAL CHARACTERISTICS:

Radius of Neptune:	_____ km, _____ % of Earth
Mass of Neptune:	
Angular size in arc minutes and seconds as seen from Earth:	
Orbit size (mean distance from the Sun):	
Sidereal day (period of rotation):	
Solar day (noon to noon on surface):	
Length of year:	

Complete the following table of Neptune's observational characteristics:*

OBSERVATIONAL CHARACTERISTICS:

Date, time:	
Azimuth, altitude:	
Currently in which constellation:	
Current apparent magnitude:	
Max possible magnitude as seen from Earth:	
Current distance from Earth:	
Distance at conjunction (aligned with the Sun):	
Distance at opposition (opposite the Sun):	

*For *constellation, azimuth, and altitude,* see POSITION IN SKY; for *distance from Earth,* see POSITION IN SPACE. For everything else, see OTHER DATA. To compute *distance at conjunction,* take *orbit size* and add Earth's distance from the Sun of 1 AU. To compute *distance at opposition,* take *orbit size* and subtract Earth's distance from Sun of 1 AU.

PART 2: NEPTUNE'S ORBITAL CHARACTERISTICS

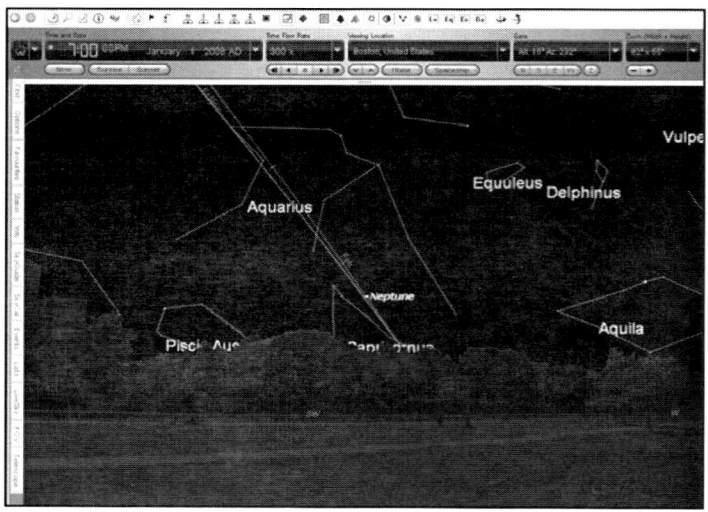

Use the **zoom control** at the far right of the **control panel** to zoom back out to full-scale view. Right click (control click for Mac) on Neptune and select **orbit**. This shows Neptune's orbital path as seen from Earth. You will need to click the **horizon** button on the **button bar** to get rid of the Earth's horizon should it interfere with your view of Neptune's orbit. You should also click the **daylight** button to turn off daylight. You should still be locked on to Neptune. If not, right click (control click for Mac) on the planet Neptune, and then select **center**. To maintain the proper perspective, from the **menu**, select **view/ecliptic guides**, then **the ecliptic**.

Start with today's date by entering it in the **date & time** field of the **control panel**, or by selecting **now** from the drop-down menu to the right of the **date & time** field. Note what constellation Neptune is in, then select a **time speed** of **1 day** and click the PLAY **time mode** button. If you need to slow down or speed up, adjust the **time speed** fields as needed. You should be able to see Neptune locked in the center of your field of view, but moving across the star background, and you should be able to complete a full revolution in less than a minute.

Click the STOP **time mode** button when Neptune has returned to the same constellation and note the date. Use this information to complete the following table.

VIEWING NEPTUNE'S RELATIVE POSITION:

Constellation Neptune is located in today:	
Today's date (corresponding to location of Neptune given above):	
Date when Neptune is again located in same constellation:	
Length of time it takes Neptune to return to the same place in the sky (subtract the dates given above):	

*Note that since Neptune is relatively far from the Sun as compared to the Earth, this value will be close to Neptune's sidereal period (year), though there will be a difference since we are looking at Neptune's position relative to the Earth, which is itself in an orbit around the Sun.

PART 3: FLYING TO NEPTUNE

Let's see what a day would be like on Neptune (technically referred to as a *solar day*). The easiest way to experience a solar day is to watch a sunset, take note of the date and time, then watch another consecutive sunset, and see what the time difference between them is.*

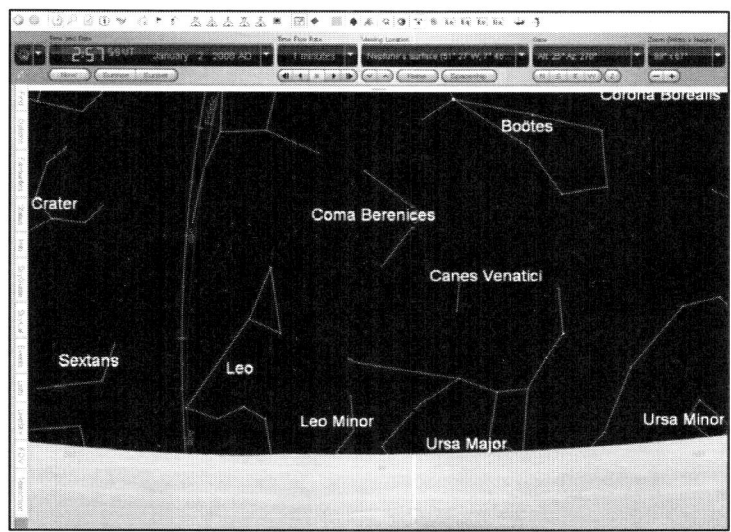

Right click (control click for Mac) on Neptune and select **go there**. You can animate the journey by first unchecking the **only animate intra planet changes** box under **file / preferences / responsiveness** (also, be sure to have your horizon turned on to see the photorealistic surface panorama). Click the **W viewing direction** button on the **button bar**, or simply hit the W key on the keyboard. Select a **time speed** of 1 **minute**. Click the PLAY **time mode** button, and click STOP when the Sun is near setting.** Use the FORWARD STEP and BACKWARD STEP **time mode** buttons until you see the Sun just starting to set. You may then need to select a smaller unit of time until the Sun is just touching the horizon (Neptune has no solid surface, so it's really the cloud tops we are viewing). Note the date and time, then continue on to the next sunset and calculate the time difference between them.

Date and time of first sunset on Neptune:	
Date and time of second sunset on Neptune:	
Subtract to get length of solar day:	

*For even greater accuracy, instead of using sunset, you should use solar noon. To find solar noon, use the LOCAL **coordinate system** button to align the Sun with the local meridian line, the Sun's highest point in the sky for that day. Starry Night also gives the time of solar noon in the **date & time** drop-down list on the **control panel**.

If you are unable to see the Sun to the west during sunset, first use the **date & time drop-down menu to select **sunset**, then click the check box to the left of the Sun in the FIND **side pane** to label the Sun, then pan to the left (SW) or right (NW) until you see it. If you still have trouble locating the Sun, right click (control click for Mac) on the Sun in the FIND **side pane** and select **centre**. You will need to unlock the Sun before continuing with this activity. The easiest way to "unlock" is to simply "grab" the sky by clicking and holding the left mouse button and then moving the mouse a little bit in any direction.

The solar day is not the same as the sidereal rotation period of a planet. This is because the solar day takes into account both the rotation of the planet and the revolution of the planet around the Sun. Your calculation of a solar day is, therefore, a little different than the sidereal day found earlier in this activity from the INFO **side pane**. How do we determine the revolutionary period, or a planet's "year?"

Go to the solar system view. From the **menu**, select **favourites/solar system/outer planets/** and select **outer solar system**. Click the STOP **time mode** button, then select **now** from the **date & time** drop-down menu. Choose a **time speed** of **1 day**, click the PLAY **time mode** button, then note the time it takes for the planet to return to its original position. You may find it easiest to align the planet to the farthest left or right before starting. Complete the chart below.

Date of Neptune at start:	
Date of Neptune after one revolution:	
Subtract to get length of year:	

Activity 16—THE MOONS of NEPTUNE

These activities are designed to work with the Starry Night software that comes with your text, from any home location you choose, and with the current date and time, unless indicated otherwise. You may always revert to factory default settings by clicking FILE/ preferences, then selecting factory defaults as needed. You may also undo a command or series of commands on the PC by clicking the back button at the top left of the button bar. You should refer to the key given at the beginning of this booklet for clarification of "on screen" buttons, controls, and functions. PC button bar items can all be accessed through the menu. "Right click" on the PC is equivalent to "control click" on the Mac. All activities assume that OpenGL graphics capabilities are enabled on your computer.

PART 1: FINDING NEPTUNE

You should begin this activity at sunset. An easy way to do this is to click the drop-down menu to the right of the **date & time** field on the **control panel**, and select **sunset**. Look toward the west by clicking the **W viewing direction** button located on the **button bar** across the top of your screen, or by simply keying in the letter W (Mac users should refer to the button bar commands given at the beginning of this booklet). The screen will pan toward the west. Select a playing speed of **300×** normal time by clicking the drop-down menu at the right of the **time speed** field. Click the STOP **time mode** button when the Sun has set, the stars have come out, and dusk is almost over. Then click on the **constellations** button to show the constellations.

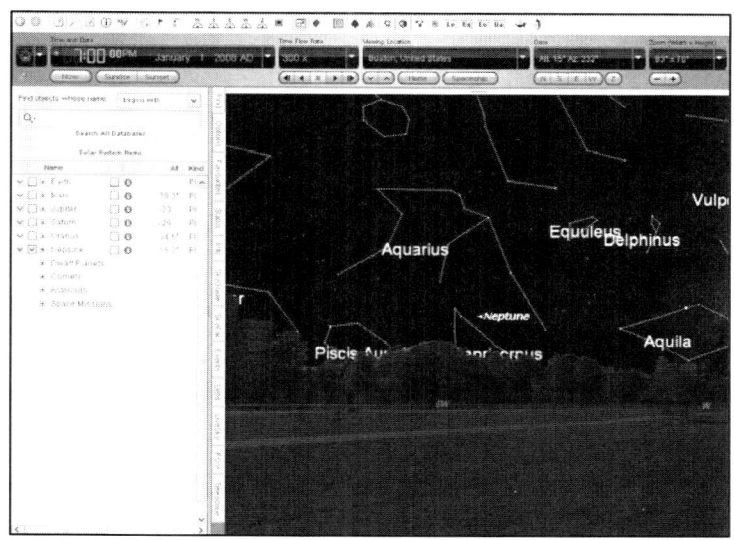

Click the FIND tab on the left **side pane**. A list of planets should appear. Those that are highlighted are currently up in your evening sky. Those that are not highlighted are not up in the sky at this time. We wish to find the planet Neptune. If Neptune is highlighted (see footnote if not highlighted)*, double-click on it or you may right click (control click for Mac) and select **centre**. This will pan the screen and center Neptune. You can now zoom in on Neptune either by using the **zoom control** at the far right of the **control panel**, or by right clicking (control clicking for Mac) on the highlighted listing in the FIND **side pane** and selecting **magnify**.

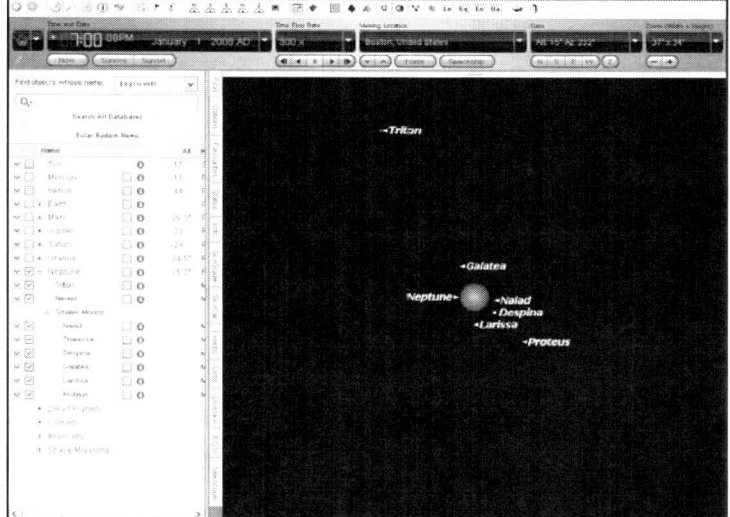

*If Neptune is not currently highlighted, you will need to move time forward to a time when Neptune will rise in the east. Start by looking toward the east by clicking the **E viewing direction** button located on the **button bar** across the top of your screen. Select a playing speed of 300× or 3000× normal time by clicking the drop-down menu at the right of the **time speed** field. Click the STOP **time mode** button when the planet Neptune becomes highlighted in the FIND **side pane** listing. It may be that Neptune is up during the day. If the Sun rises before Neptune does, then click the **daylight** button on the **button bar**. This will keep the sky dark so you can see the stars and constellations. Once Neptune has risen, double-click on it. This will pan the screen and center Neptune. You can now zoom in on Neptune either by using the **zoom control** at the far right of the **control panel**, or by right clicking (control clicking for Mac) on the highlighted listing in the FIND **side pane** and selecting **magnify**.

Click the information icon **(i)** in the **side pane** to read a short description of the planet Neptune. For even more information, select the INFO tab on the left **side pane** and click the plus sign (gray arrow for Mac) to expand the different information categories.

PART 2: NEPTUNE'S MOON TRITON

We wish to take a closer look at Neptune's largest moon, Triton.

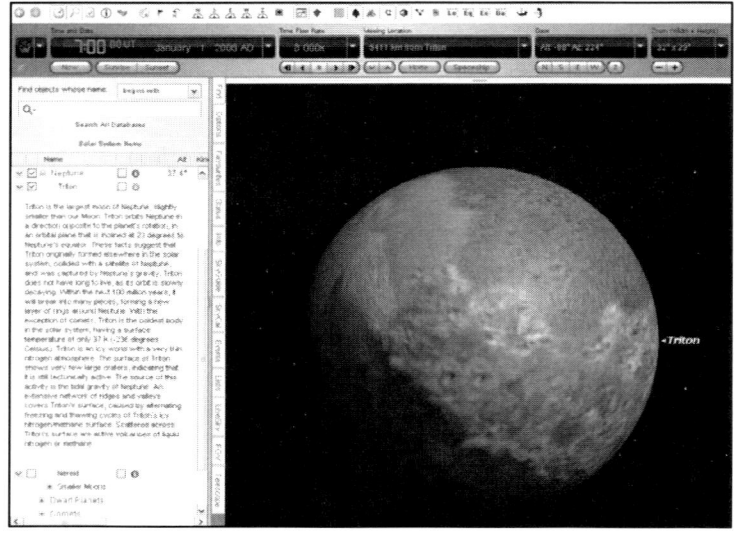

Click the plus sign (gray arrow for Mac) next to **Neptune** in the FIND **side pane**. This will list Neptune's moons. Click the first check box for **Triton** to label it on your screen. In order to see both the planet and the moon, you may need to pan out by using the **zoom control** at the far right of the **control panel**. Keep in mind that Triton may be out of sight behind the planet itself. In that case, use the **3000× time mode** setting on the **control panel** to move time forward so that Triton is clearly visible. Remember that as you move time forward, you may need to click the **horizon** and **daylight** buttons on the **button bar** as needed.

Let's take a closer look at Triton. Double click on **Triton** in the list. This will center Triton on your screen. Use the **zoom control** to get a close up view.

[OPTIONAL] If you would like to get an even closer view beyond the limits of the zoom function, right click (control click for Mac) on the moon and select **go there**. You are now on the surface (to see the surface, make sure your horizon is turned on). Now launch yourself up above the moon by clicking the **location above surface of planet** up arrow on the **control panel**. Lock on to Triton by double-clicking on **Triton** in the FIND **side pane**, then use the **zoom control** as needed to see the moon full screen. When done, be sure to click the **back** button on the **button bar** a few times to return to an earthbound perspective for the rest of the assignment.

Click the information icon **(i)** in the **side pane** to read a short description of this moon. Then, select the INFO tab on the left **side pane** and click the plus sign (gray arrow for Mac) to expand the different information categories.

Complete the following table:

Compare to Earth's Moon with radius = 1737 km, and mass = 7.35×10^{22} kg = 0.0123 Earth masses.

Neptune's radius is 24763 km, Neptune's mass is 17.221 Earth masses, and Earth's mass = 5.974×10^{24} kg.

ATTRIBUTE	TRITON
Radius (in km):	
…in multiples of Earth's Moon radius:*	
Mass (in Earth masses):	
…in multiples of Earth's Moon mass:	
…as ratio of Neptune's mass:	
Orbit size (in km):	
…in multiples of Neptune's radius:	
Sidereal day (one rotation):	
Solar day (noon to noon):	
Year (to orbit around Neptune):	

*The info given for *radius* is in km. To compute the *multiple of Earth's moon radius* take *radius* in km and divide by *Earth's moon radius* as given above. Follow the same procedure for *mass* and *orbit size*.

Note how small the ratio of Triton's mass is as compared to the mass of Neptune itself. Its mass ratio is very small, with little gravitational effect on Neptune. In contrast, Earth's moon-to-planet ratio is one of the largest in the solar system at 0.0123 (or 1.23%).

PART 3: SATTELITE ORBITAL CHARACTERISTICS

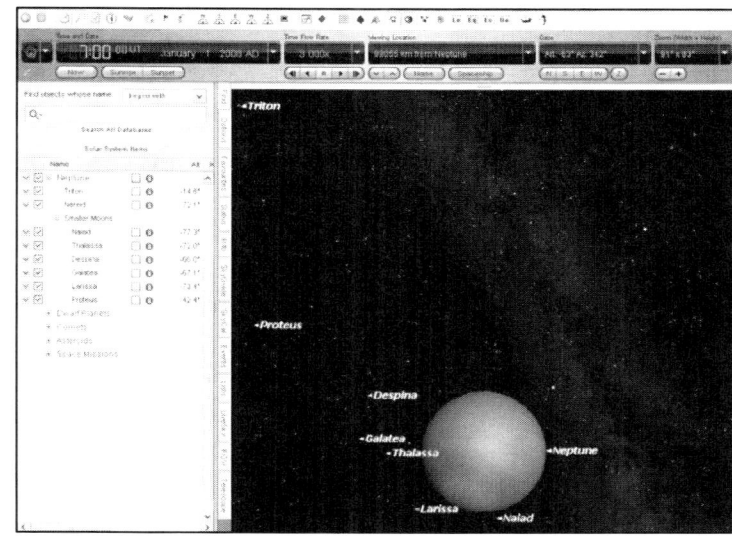

Zoom back out so you can see Neptune and its moons once again. Click the **labels** button on the **button bar** to see the names of other moons. Double click on **Neptune** in the FIND **side pane** to center on Neptune and zoom in and out a bit to see how many moons you can observe. Select the **3000× time speed** on the **control panel** and watch the moons orbit Neptune. Click the **daylight** and **horizon** buttons as needed. For best results, so that the smaller moons will be displayed, Neptune should cover about a third of your screen. Watch the motion for a bit. The tilting is due to our local perspective. You can remove this effect by selecting **view/ecliptic guides**, then **the ecliptic**. Why do the moons move from side-to-side rather than in circles around Neptune? This also has to do with perspective. Since Neptune's satellite plane lies along the ecliptic, the plane of our solar system, we can only observe Neptune from the side.

[OPTIONAL] You can use Starry Night to observe the moons from above Neptune's polar region. First use Starry Night to go to Neptune, then change your position to Neptune's north pole. Next, launch yourself straight up until you can view the moons from above. Now when you click the **play** button, you will see the moons orbiting in nearly circular orbits. When done, click the **back** button repeatedly until you have returned to your original perspective from Earth.

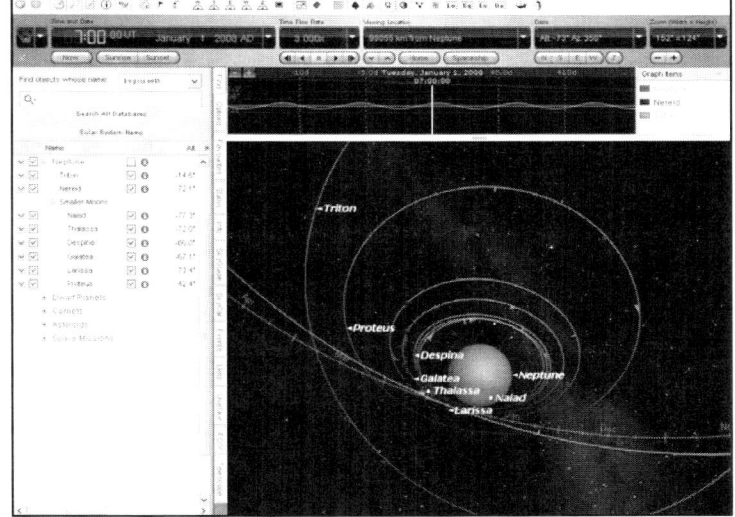

Let's take a closer look at satellite orbits. To better visualize the satellite orbital motion, click both the first and second set of check boxes in the FIND **side pane** for all of Neptune's moons to label and trace out their respective orbits around Neptune. Again, you may need to adjust the **zoom controls** to better view the orbital plane. You should notice that the two outermost moons deviate significantly from the orbital plane defined by the rest of the moons. One of these is Triton, the largest, orbiting in retrograde. The other is Nereid, which finds itself in a highly elliptical orbit with, Neptune at one focus.

To get a feel for how fast these moons move around in their orbit, slowly move time forward using the **300×** or **3000× time speeds** and observe the orbital motion as it would appear as seen from Earth.

To see some different viewing perspectives of the moons, right click (control click for Mac) on Neptune and select **graph elongation of moons**. The graph shows the angular separation between numerous moons with Neptune. Triton and Nereid are graphed, but Proteus is not.

To add Proteus to the graph, right click (control click for Mac) on **Proteus** either in the FIND **side pane** or on the screen and select **start graphing**. Similarly, you can remove moons from the graph by right clicking (control clicking for Mac) on them and selecting **stop graphing**. Nereid has such a long orbital period that you will need to click on the minus sign at the upper left of the graph to compress the graph so you can observe a full orbital cycle (or better yet, multiple orbital cycles).

You can move time forward with the **time speed** fields on the **control panel**, or you can simply grab the graph by clicking down on it with the mouse then dragging to the left or right while holding the mouse button down. You can also expand the horizontal resolution of the graph by clicking on the plus and minus indicators at the top left of the graph. To increase the vertical resolution, grab the bottom of the graph with the mouse and pull down.* By moving the graph back and forth to look for different alignments, you can answer the following questions.

SATELLITE ORBITAL CHARACTERISTICS:

Time for a complete cycle of Triton:	
Maximum elongation for Triton (see vertical scale)**:	
Time for a complete cycle of Nereid:	
Maximum elongation for Nereid (see vertical scale)**:	
Elongation for Nereid at other side of orbit (see vertical scale):	

*You may need to select **stop graphing** for Nereid in order to increase the vertical resolution of Titan and the other moons.

**You can check and improve on the precision of the maximum elongation values by centering your pointer on Neptune, then clicking down and dragging the pointer to the farthest extent of the moon's orbit (make sure the orbit trace is on). The angular separation will be displayed.

Activity 17—PLUTO

*These activities are designed to work with the Starry Night software that comes with your text, from any home location you choose, and with the current date and time, unless indicated otherwise. You may always revert to factory default settings by clicking **FILE/ preferences**, then selecting **factory defaults** as needed. You may also undo a command or series of commands on the PC by clicking the **back** button at the top left of the **button bar**. You should refer to the key given at the beginning of this booklet for clarification of "on screen" buttons, controls, and functions. PC **button bar** items can all be accessed through the **menu**. "Right click" on the PC is equivalent to "control click" on the Mac. All activities assume that OpenGL graphics capabilities are enabled on your computer.*

PART 1: FINDING PLUTO

You should begin this activity at sunset. An easy way to do this is to click the drop-down menu to the right of the **date & time** field on the **control panel**, and select **sunset**. Look toward the west by clicking the **W viewing direction** button located on the **button bar** across the top of your screen, or by simply keying in the letter W (Mac users should refer to the button bar commands given at the beginning of this booklet). The screen will pan toward the west. Select a playing speed of **300×** normal time by clicking the drop-down menu at the right of the **time speed** field. Click the STOP **time mode** button when the Sun has set, the stars have come out, and dusk is almost over. Then click on the **constellations** button to show the constellations.

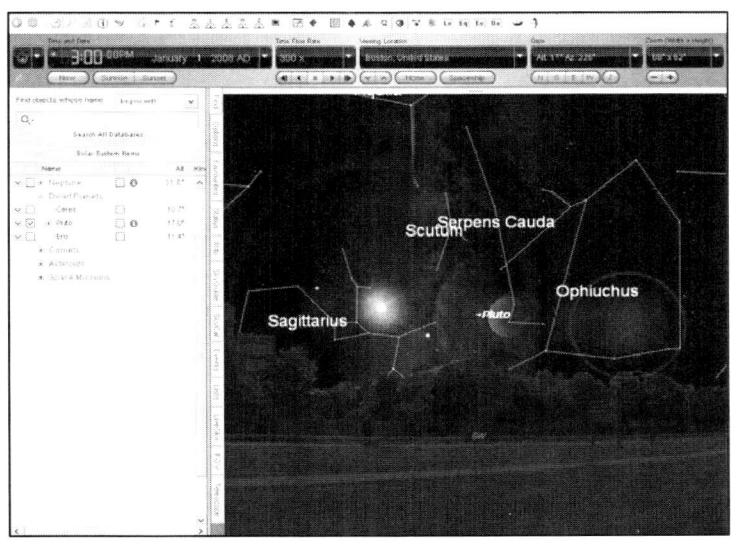

Click the FIND tab on the left **side pane**. A list of planets and dwarf planets should appear. Those that are highlighted (see footnote if not highlighted)* are currently up in your evening sky. Those that are not highlighted are not up in the sky at this time. We wish to find Pluto. If Pluto is highlighted, double-click or right click on it (control click for Mac) and select **centre**. This will pan the screen and center Pluto. You can now zoom in on Pluto either by using the **zoom control** at the far right of the **control panel**, or by right clicking (control clicking for Mac) on the highlighted listing in the FIND **side pane** and selecting **magnify**.

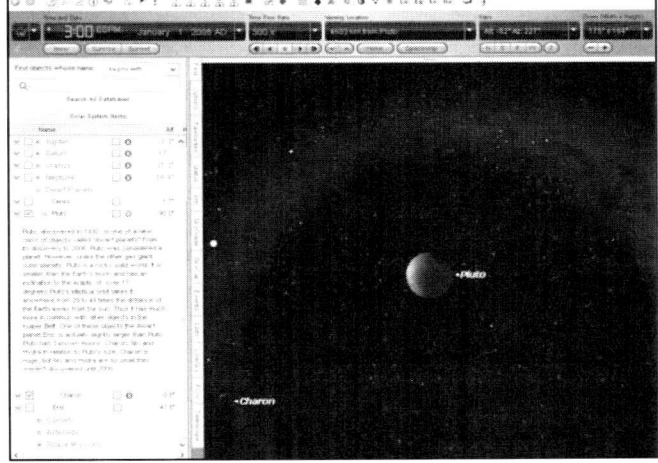

*If Pluto is not currently highlighted, you will need to move time forward to a time when Pluto will rise in the east. Start by looking toward the east by clicking the **E viewing direction** button located on the **button bar** across the top of your screen. Select a playing speed of **300×** or **3000×** normal time by clicking the drop-down menu at the right of the **time speed** field. Click the STOP **time mode** button when the planet Pluto becomes highlighted in the FIND **side pane** listing. It may be that Pluto is up during the day. If the Sun rises before Pluto does, then click the **daylight** button on the **button bar**. This will keep the sky dark so you can see the stars and constellations. Once Pluto has risen, double-click on it. This will pan the screen and center Pluto. You can now zoom in on Pluto either by using the **zoom control** at the far right of the **control panel**, or by right clicking (control clicking for Mac) on the highlighted listing in the FIND **side pane** and selecting **magnify**.

Click the information icon **(i)** in the **side pane** to read a short description of the planet Pluto. Next, select the INFO tab on the left **side pane** and click the plus sign (gray arrow for Mac) to expand the different information categories. Use the information in the OTHER DATA section to complete the following table of Pluto's physical characteristics and compare to Earth.

(You will need to know Earth's radius = 6378 km. To compute percent, divide one value by the other and multiply by 100.)

PHYSICAL CHARACTERISTICS:

Radius of Pluto:	_____ km, _____ % of Earth
Mass of Pluto:	
Angular size in arc minutes and seconds as seen from Earth:	
Orbit size (mean distance from Sun):	
Sidereal day (period of rotation):	
Solar day (noon to noon on surface):	
Length of year:	

Complete the following table of Pluto's observational characteristics:*

OBSERVATIONAL CHARACTERISTICS:

Date, time:	
Azimuth, altitude:	
Currently in which constellation:	
Current apparent magnitude:	
Max possible magnitude as seen from Earth:	
Current distance from Earth:	
Distance at conjunction (aligned with the Sun):	
Distance at opposition (opposite the Sun):	

*For *constellation, azimuth, and altitude,* see POSITION IN SKY; for *distance from Earth,* see POSITION IN SPACE. For everything else, see OTHER DATA. To compute *distance at conjunction,* take *orbit size* and add Earth's distance from the Sun of 1 AU. To compute *distance at opposition,* take *orbit size* and subtract Earth's distance from Sun of 1 AU.

PART 2: PLUTO'S ORBITAL CHARACTERISTICS

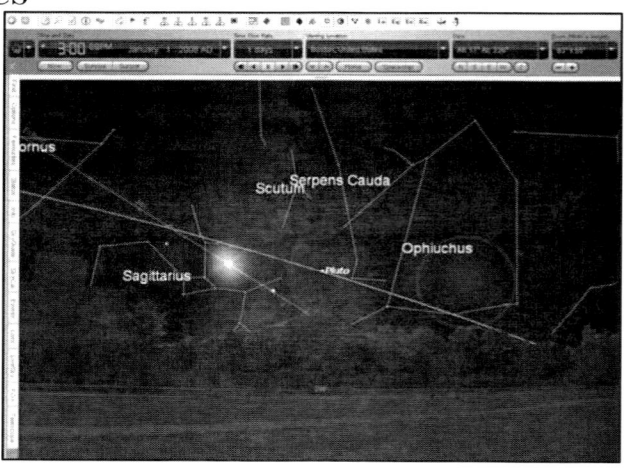

Use the **zoom control** at the far right of the **control panel** to zoom back out to full-scale view. Right click (control click for Mac) on Pluto and select **orbit**. This shows Pluto's orbital path as seen from Earth. You will need to click the **horizon** button on the **button bar** to get rid of the Earth's horizon should it interfere with your view of Pluto's orbit. You should also click the **daylight** button to turn off daylight. You should still be locked on to Pluto. If not, right click (control click for Mac) on Pluto, and then select **center**. To maintain the proper perspective, from the **menu** select **view/ecliptic guides**, then **the ecliptic**.

Start with today's date by entering it in the **date & time** field of the **control panel**, or by selecting **now** from the drop-down menu to the right of the **date & time** field. Note what constellation Pluto is in, then select a **time speed** of **1 day** and click the PLAY **time mode** button. If you need to slow down or speed up, adjust the **time speed** fields as needed. You should be able to see Pluto locked in the center of your field of view, but moving across the star background, and you should be able to complete a full revolution in less than a minute. Click the STOP **time mode** button when Pluto has returned to the same constellation and note the date. Use this information to complete the table below.

VIEWING PLUTO'S RELATIVE POSITION:

Constellation Pluto is located in today:	
Today's date (corresponding to location of Pluto given above):	
Date when Pluto is again located in same constellation:	
Length of time it takes Pluto to return to the same place in the sky (subtract the dates given above):	

*Note that since Pluto is relatively far from the Sun as compared to the Earth, this value will be close to Pluto's sidereal period (year), though there will be a difference since we are looking at Pluto's position relative to the Earth, which is itself in an orbit around the Sun.

PART 3: FLYING TO PLUTO

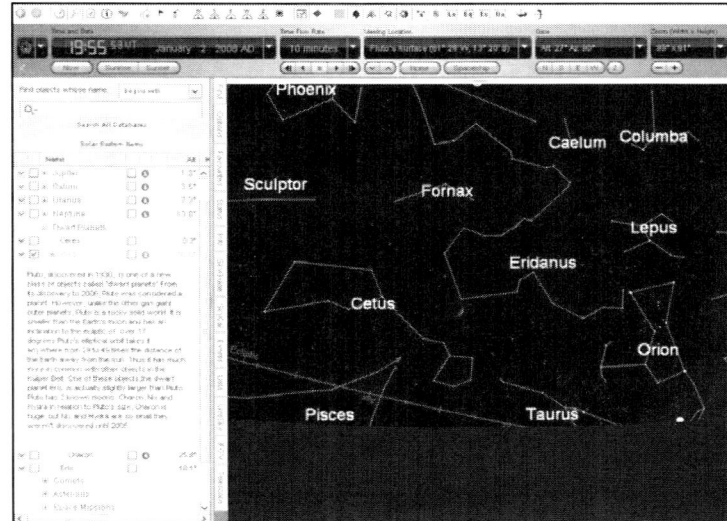

Let's see what a day would be like on Pluto (technically referred to as a *solar day*). The easiest way to experience a solar day is to watch a sunset, take note of the date and time, then watch another consecutive sunset, and see what the time difference between them is.*

Right click (control click for Mac) on Pluto and select **go there**. You can animate the journey by first unchecking the **only animate intra planet changes** box under **file/preferences/ responsiveness** (also, be sure to have your horizon turned on to see the photorealistic surface panorama). Click the **E viewing direction** button on the **button bar**, or simply hit the E key on the keyboard (note that since Pluto rotates in retrograde, sunset will be to the East). Select a **time speed** of **10 minutes**. Click the PLAY **time mode** button, and click STOP when the Sun is near setting.** Use the FORWARD STEP and BACKWARD STEP **time mode** buttons until you see the Sun just starting to set. You may then need to select a smaller unit of time until the Sun is just touching the horizon. Note the date and time, then continue on to the next sunset and calculate the time difference between them.

Date and time of first sunset on Pluto:	
Date and time of second sunset on Pluto:	
Subtract to get length of Solar Day:	

*For even greater accuracy, instead of using sunset, you should use solar noon. To find solar noon, use the LOCAL **coordinate system** button to align the Sun with the local meridian line, the Sun's highest point in the sky for that day. Starry Night also gives the time of solar noon in the **date & time** drop-down list on the **control panel**.

If you are unable to see the Sun to the east during sunset, first use the **date & time drop-down menu to select **sunset**, then click the check box to the left of the Sun in the FIND **side pane** to label the Sun, then pan to the left (NE) or right (SE) until you see it. If you still have trouble locating the Sun, right click (control click for Mac) on the Sun in the FIND **side pane** and select **centre**. You will need to unlock the Sun before continuing with this activity. The easiest way to "unlock" is to simply "grab" the sky by clicking and holding the left mouse button, and then moving the mouse a little bit in any direction.

The solar day is not the same as the sidereal rotation period of a planet (or Pluto). This is because the solar day takes into account both the rotation of the dwarf planet and the revolution of the dwarf planet around the Sun. Your calculation of a solar day is, therefore, different than the sidereal day found earlier in this activity from the INFO **side pane**.

How do we determine the revolutionary period, or "year", of Pluto?

Go to the solar system view. From the **menu**, select **favourites/solar system/outer planets/** and select **pluto and beyond**. Click the STOP **time mode** button, then select **now** from the **date & time** drop-down menu. Choose a **time speed** of **1 day**, click the PLAY **time mode** button, then note the time it takes for the planet to return to its original position (you may need to increase the number of days depending on the speed of your computer). You may find it easiest to align the planet to the farthest left or right before starting. Complete the chart below.

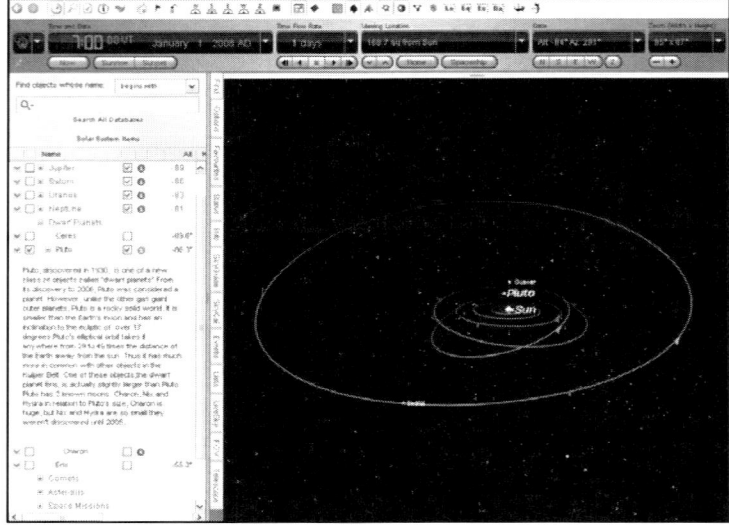

Date of Pluto at start:	
Date of Pluto after one revolution:	
Subtract to get length of year:	

Activity 18—PLUTO and CHARON

These activities are designed to work with the Starry Night software that comes with your text, from any home location you choose, and with the current date and time, unless indicated otherwise. You may always revert to factory default settings by clicking FILE/preferences, then selecting factory defaults as needed. You may also undo a command or series of commands on the PC by clicking the back button at the top left of the button bar. You should refer to the key given at the beginning of this booklet for clarification of "on screen" buttons, controls, and functions. PC button bar items can all be accessed through the menu. "Right click" on the PC is equivalent to "control click" on the Mac. All activities assume that OpenGL graphics capabilities are enabled on your computer.

PART 1: FINDING PLUTO

You should begin this activity at sunset. An easy way to do this is to click the drop-down menu to the right of the **date & time** field on the **control panel**, and select **sunset**. Look toward the west by clicking the **W viewing direction** button located on the **button bar** across the top of your screen, or by simply keying in the letter W (Mac users should refer to the button bar commands given at the beginning of this booklet). The screen will pan toward the west. Select a playing speed of **300×** normal time by clicking the drop-down menu at the right of the **time speed** field. Click the STOP **time mode** button when the Sun has set, the stars have come out, and dusk is almost over. Then click on the **constellations** button to show the constellations.

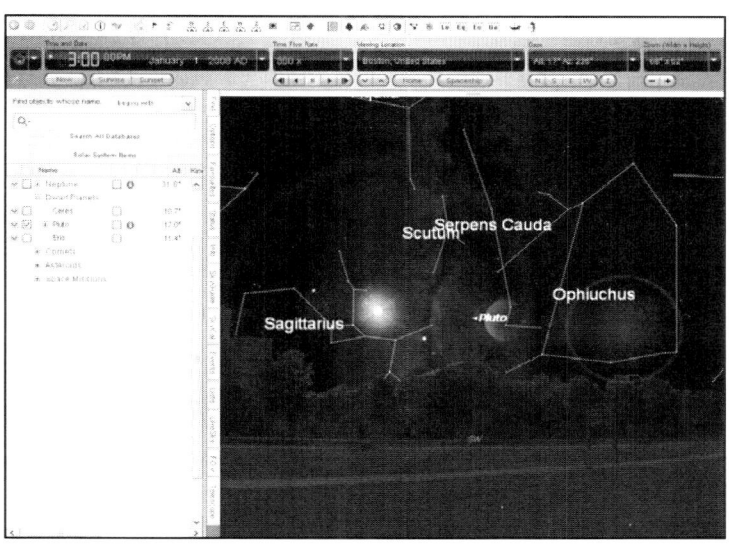

Click the FIND tab on the left **side pane**. A list of planets and dwarf planets should appear. Those that are highlighted are currently up in your evening sky. Those that are not highlighted are not up in the sky at this time. We wish to find Pluto. If Pluto is highlighted (see footnote if not highlighted)*, double-click or right click on it (control click for Mac) and select **centre**. This will pan the screen and center Pluto. You can now zoom in on Pluto either by using the **zoom control** at the far right of the **control panel**, or by right clicking (control clicking for Mac) on the highlighted listing in the FIND **side pane**, and selecting **magnify**.

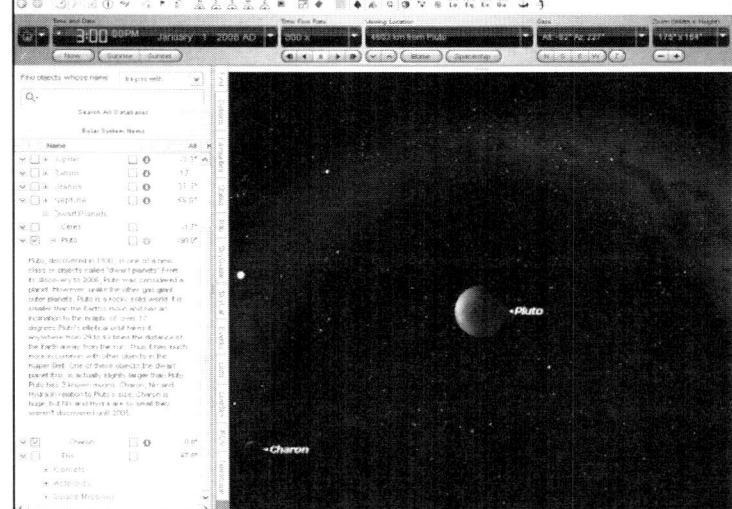

*If Pluto is not currently highlighted, you will need to move time forward to a time when Pluto will rise in the east. Start by looking toward the east by clicking the **E viewing direction** button located on the **button bar** across the top of your screen. Select a playing speed of **300×** or **3000×** normal time by clicking the drop-down menu at the right of the **time speed** field. Click the STOP **time mode** button when the planet Pluto becomes highlighted in the FIND **side pane** listing. It may be that Pluto is up during the day. If the Sun rises before Pluto does, then click the **daylight** button on the **button bar**. This will keep the sky dark so you can see the stars and constellations. Once Pluto has risen, double-click on it. This will pan the screen and center Pluto. You can now zoom in on Pluto either by using the **zoom control** at the far right of the **control panel**, or by right clicking (control clicking for Mac) on the highlighted listing in the FIND **side pane** and selecting **magnify**.

Click the information icon (**i**) in the **side pane** to read a short description of Pluto. For even more information, select the INFO tab on the left **side pane** and click the plus sign (gray arrow for Mac) to expand the different information categories.

PART 2: PLUTO and CHARON

We wish to take a closer look at Pluto's largest known "moon", Charon.

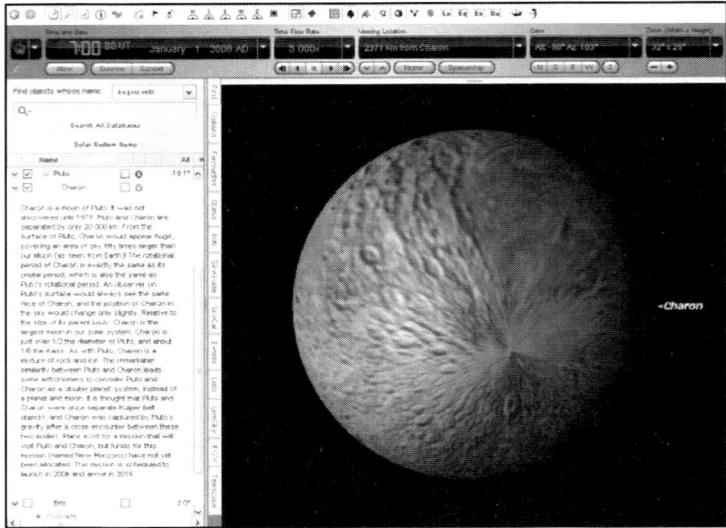

Click the plus sign (gray arrow for Mac) next to **Pluto** in the FIND **side pane**. This will list the moon Charon. Click the first check box for **Charon** to label it on your screen. In order to see both the planet and the moon, you may need to pan out by using the **zoom control** at the far right of the **control panel**. Keep in mind that Charon may be out of sight behind the planet itself. In that case, use the **30000× time mode** setting on the **control panel** to move time forward so that Charon is clearly visible. Remember that as you move time forward, you should click the **horizon** and **daylight** buttons on the **button bar** as needed.

Let's take a closer look at Charon. Double click on **Charon** in the list. This will center Charon on your screen. Use the **zoom control** to get a close up view.

[OPTIONAL] If you would like to get an even closer view beyond the limits of the zoom function, right click (control click for Mac) on the moon and select **go there**. You are now on the surface (to see the surface, make sure your horizon is turned on). Now launch yourself up above the moon by clicking the **location above surface of planet** up arrow on the **control panel**. Lock onto Charon by double-clicking on **Charon** in the FIND **side pane**, then use the **zoom control** as needed to see the moon full screen. Note that this is not a completely accurate image of Charon. No satellite imagery exists of Pluto or Charon, but the designer of this spherical representation may have used Hubble Space Telescope images in an effort to incorporate potential surface features and characterizations. When done, be sure to click the **back** button on the **button bar** a few times to return to an earthbound perspective for the rest of the assignment.

Click the information icon **(i)** in the **side pane** to read a short description of this moon. Then, select the INFO tab on the left **side pane** and click the plus sign (gray arrow for Mac) to expand the different information categories.

Complete the following table:

Compare to Earth's moon with radius = 1737 km, and mass = 7.35×10^{22} kg = 0.0123 Earth masses.

Pluto's radius is 1150 km, Pluto's mass is 0.1109 Earth masses, and Earth's mass = 5.974×10^{24} kg.

ATTRIBUTE	CHARON
Radius (in km):	
…in multiples of Earth's Moon radius:*	
Mass (in Earth Masses):	
…in multiples of Earth's Moon mass:	
…as ratio of Pluto's mass:	
Orbit Size (in km):	
…in multiples of Pluto's radius:	
Sidereal day (one rotation):	
Solar day (noon to noon):	
Year (to orbit around Pluto):	

*The info given for *radius* is in km. To compute the *multiple of Earth's moon radius* take *radius* in km and divide by *Earth's moon radius* as given above. Follow the same procedure for *mass* and *orbit size*.

Note how large the ratio of Charon's mass is as compared to the mass of Pluto itself. Its moon-to-(dwarf) planet mass ratio is the largest in the solar system, with significant gravitational effects on Pluto. Compare this ratio with the very small ratios of other moons in the solar system. Also compare Charon's mass-to-(dwarf) planet ratio with Earth's moon-to-planet ratio, the second largest in the solar system at 0.0123 (or 1.23%). You can see why many scientists consider Pluto and Charon a double planet (or dwarf planet) system.

PART 3: SATTELITE ORBITAL CHARACTERISTICS

Zoom back out so you can see Pluto and Charon again. Select the **30000× time speed** on the **control panel** and watch the moon orbit Pluto. Click the **daylight** and **horizon** buttons as needed. Watch the motion for a bit. The tilting is due to our local perspective. You can remove this effect by selecting **view/ecliptic guides**, then **the ecliptic.**

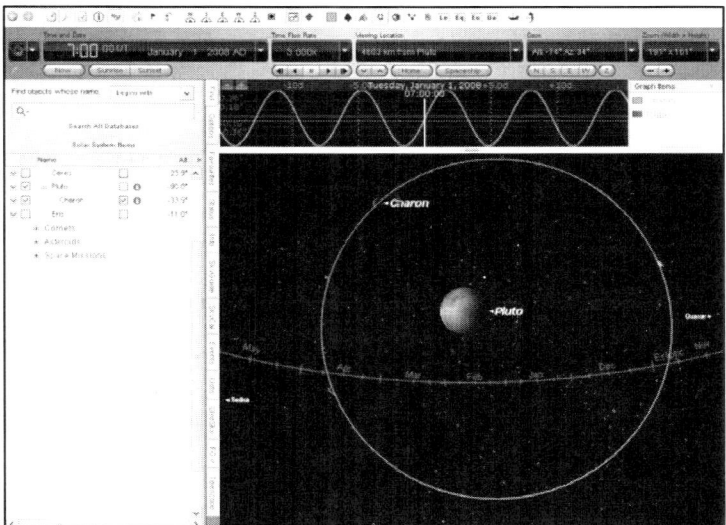

Let's take a closer look at Charon's orbit. To better visualize the satellite orbital motion, click both the first and second set of check boxes in the FIND **side pane** for **Charon** to label and trace out its orbit around Pluto. Again, you may need to adjust the **zoom controls** to better view the orbital plane. To get a feel for how fast Charon moves around in its orbit, slowly move time forward using the **3000×** or **30000× Time Speed** and observe the orbital motion as it would appear as seen from Earth.

To see a different viewing perspective of Charon, right click (control click for Mac) on Pluto and select **graph elongation of moons**. The graph shows the angular separation between Charon with Pluto. You can move time forward with the **time speed** fields on the **control panel**, or you can simply grab the graph by clicking down on it with the mouse, then dragging to the left or right while holding the mouse button down. You can also expand the horizontal resolution of the graph by clicking on the plus and minus indicators at the top left of the graph. To increase the vertical resolution, grab the bottom of the graph with the mouse and pull down. By moving the graph back and forth, you can answer the following questions.

SATELLITE ORBITAL CHARACTERISTICS:

Time for a complete cycle of Charon	
Maximum elongation for Charon (see vertical scale)*	

*You can check and improve on the precision of the maximum elongation value by centering your pointer on Pluto, then clicking down and dragging the pointer to the farthest extent of the moon's orbit (make sure the orbit trace is on). The angular separation will be displayed.

Activity 19—PHYSICAL PROPERTIES of STARS

*These activities are designed to work with the Starry Night software that comes with your text, from any home location you choose, and with the current date and time, unless indicated otherwise. You may always revert to factory default settings by clicking **FILE/ preferences**, then selecting **factory defaults** as needed. You may also undo a command or series of commands on the PC by clicking the **back** button at the top left of the **button bar**. You should refer to the key given at the beginning of this booklet for clarification of "on screen" buttons, controls, and functions. PC **button bar** items can all be accessed through the **menu**. "Right click" on the PC is equivalent to "control click" on the Mac. All activities assume that OpenGL graphics capabilities are enabled on your computer.*

PART 1: DIFFERENT TYPES OF STARS

The stars you see at night are all very different. There are stars (stars by themselves); variable stars (stars that vary in brightness); binary stars (pairs of stars); multiple star systems (all gravitationally bound together); main sequence stars; red and blue giants; red and blue supergiants; and a host of other star-like objects, brown dwarfs, white dwarfs, neutron stars, pulsars, and black holes. With so much variety in the sky, how do we know what we are looking at?

This activity will focus on stars and the physical properties of stars (such as size, temperature, color, and brightness).

PART 2: STARS OF OUR CURRENT EVENING SKY

You should begin this activity at sunset. An easy way to do this is to click the drop-down menu to the right of the **date & time** field on the **control panel**, and select **sunset**. Look toward the west by clicking the **W viewing direction** button located on the **button bar** across the top of your screen, or by simply keying in the letter W (Mac users should refer to the button bar commands given at the beginning of this booklet). The screen will pan toward the west. Select a playing speed of **300×** normal time by clicking the drop-down menu at the right of the **time speed** field. Click the stop **time mode** button when the Sun has set, the stars have come out, and dusk is almost over. Then click on the **constellations** button to show the constellations, and click the **labels** button to label stars and objects.

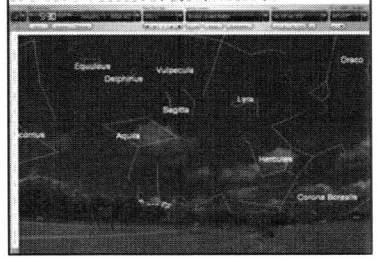
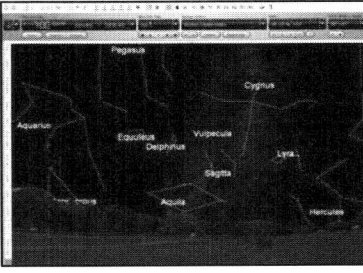
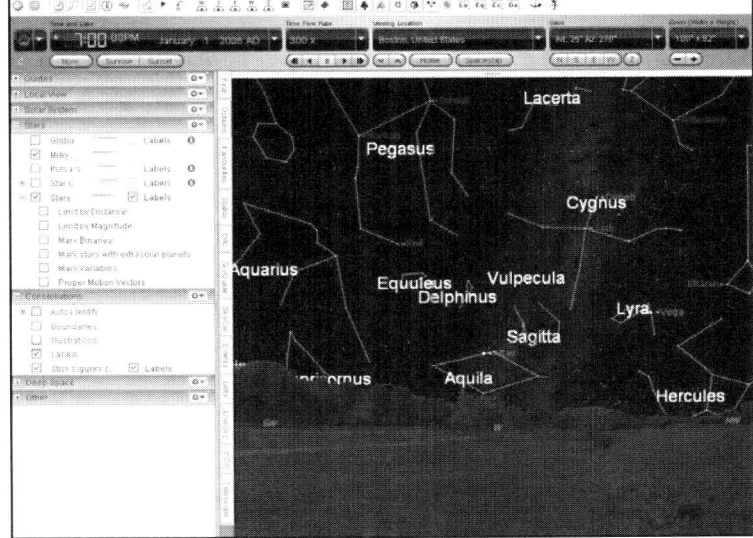

PART 3: SOME REPRESENTATIVE STARS

Let's see some examples of different types of stars; a typical field star such as Alpha Centauri (one of our nearest neighbors and about the same size as our Sun), a red giant/supergiant such as Betelgeuse (in Orion, the mighty hunter), a blue giant such as Deneb (in Cygnus, the swan), and Barnard's Star (a small, dim, nearby star that can't even be seen with the naked eye).

ALPHA CENTAURI (HIP71683)

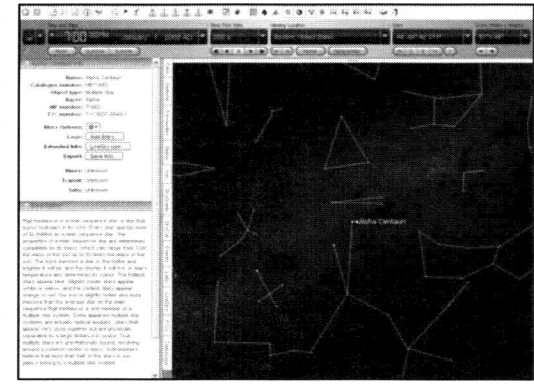

Alpha Centauri is a typical star on the main sequence. Note how this star's radius, temperature, and luminosity compare to our Sun. Such stars are fusing hydrogen to helium. This is a most efficient nuclear fuel (0.7% conversion to pure energy by Einstein's $E = mc^2$). Stars will remain on the main sequence for millions, billions, or even trillions of years. A star's mass determines how long its fuel will last. Massive stars burn fuel faster due to greater gravitational forces in their cores producing higher pressures and temperatures. Such stars may only last a few million years before consuming their fuel. This may seem paradoxical since large-mass stars have more fuel to burn. However, their greater mass does not compensate for the even greater rate of nuclear fusion. A star does not move much from its position on the main sequence throughout its hydrogen-burning phase (a slight brightening does occur, but this results in only a very slight shift off the main sequence toward the upper right).

Click the FIND tab on the left **side pane**. In the search field at the top, type in ALPHA CENTAURI. Right click (control click for Mac) on the object and select **show info**, click the plus signs (gray arrow for Mac) to expand the information layers in the side pane, then complete the following table of information.

Date, time:	
Azimuth, altitude:	
In constellation:	
Bayer designation (alpha, beta, gamma, etc.):	
Distance from Sun/observer (give units):	
Radius:	
Temperature:	
Apparent magnitude:	
Absolute magnitude:	
Luminosity:	

Provide a sketch of the constellation in which this star is located, indicating the star's location.

BETELGEUSE (HIP27989)

Betelgeuse is a red giant/supergiant star. Be sure to take note of this star's radius, temperature, and luminosity. The star is very large (many times larger than our Sun), but cooler than our Sun (which is why it appears red in our night sky). However, it is extremely bright (note how many times brighter than our Sun) due to its large size.

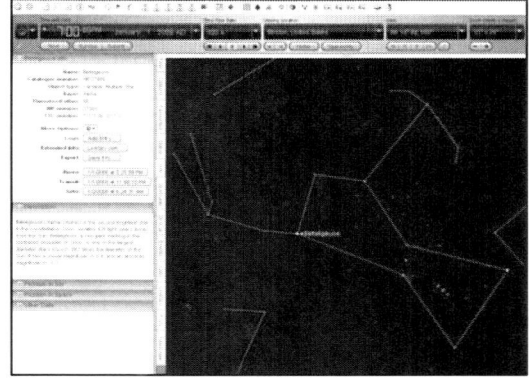

Betelgeuse is a typical red giant/supergiant. These types of stars are near the end of their stellar lives. Having run out of hydrogen fuel, this star has left the main sequence to become brighter, yet cooler and redder. It has moved toward the upper right of the H–R diagram. The star's core contracts, further heating the central core, yet expanding its outer shell. Depending on their mass, red giants/supergiants may end up as planetary nebulae (with a central white dwarf) if $M_{core} < 1.4$ solar masses; they may go supernova, resulting in pulsars (spinning neutron stars) if $M_{core} > 1.4$ solar masses; they may even become black holes (gravitationally collapsed objects whose core material after going supernova must be greater than 3 solar masses).

Click the FIND tab on the left **side pane**. In the search field at the top, type in BETELGEUSE. Right click (control click for Mac) on the object and select **show info**, click the plus signs (gray arrow for Mac) to expand the information layers in the side pane, then complete the following table of information.

Date, time:	
Azimuth, altitude:	
In constellation:	
Bayer designation (alpha, beta, gamma, etc.):	
Distance from Sun/observer (give units):	
Radius:	
Temperature:	
Apparent magnitude:	
Absolute magnitude:	
Luminosity:	

Provide a sketch of the constellation in which this star is located, indicating the exact location of the star.

DENEB (HIP102098)

Deneb is a blue giant/supergiant. Note the surface temperature of this star as compared to our Sun. Hotter temperatures imply a bluish color. Cooler temperatures imply a reddish color. Also note the radius, luminosity, and distance of this star. This star is surprisingly bright for being so very far away from us. Such tremendous power output is due to the fourth power relationship of brightness to temperature. Higher temperatures have a greater effect on brightness than do size and distance, which vary only by the square of the variable.

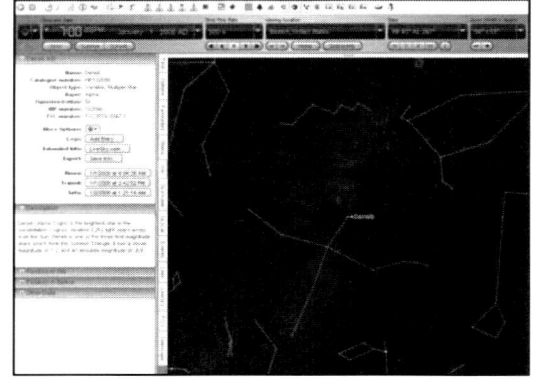

Click the FIND tab on the left **side pane**. In the search field at the top, type in DENEB. Right click (control click for Mac) on the object and select **show info**, click the plus signs (gray arrow for Mac) to expand the information layers in the side pane, then complete the following table of information.

Date, time:	
Azimuth, altitude:	
In constellation:	
Bayer designation (alpha, beta, gamma, etc.):	
Distance from Sun/observer (give units):	
Radius:	
Temperature:	
Apparent magnitude:	
Absolute magnitude:	
Luminosity:	

Provide a sketch of the constellation in which this star is located, indicating the exact location of the star.

BARNARD'S STAR (HIP71683)

Barnard's Star is a low mass star that is very dim. Note how this star's mass, radius, temperature, and luminosity compare to our Sun. This star is barely massive enough to have initiated nuclear fusion. Although this star is relatively close to us, it is all but impossible to see without the aid of high-powered telescopes and equipment.

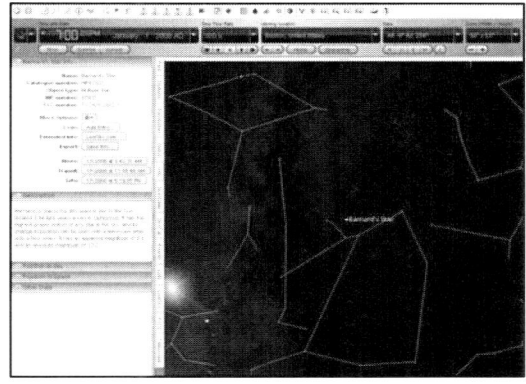

Compare this star's luminosity characteristics to that of Deneb discussed earlier. Such comparisons can serve to give us a greater appreciation of the full extent of luminosity ranges that stars can have. Another interesting and somewhat paradoxical consideration is to realize that the stars we see at night are not a representative sample of the stars in our galaxy. Although Barnard's Star is one of the closest stars to us, we can't even hope to see it with the naked eye. However, Deneb (along with other giants/supergiants) appears as one of the brightest stars in our night sky, even at thousands of light years distance. This biases our perspective of the heavens. We *think* stars are bluish-whitish. However, most stars are orangish-reddish, but just too dim to see. We think stars are bright; however, most stars are relatively dim, so dim in fact that we can't even see most of them with the naked eye. So the typical star is not as we see it in the night sky. Careful surveys are needed to get a sense of the true distribution of stellar characteristics, as seen or not seen by the naked eye.

Click the FIND tab on the left **side pane**. In the search field at the top, type in BARNARD'S STAR. Right click (control click for Mac) on the object and select **show info**, click the plus signs (gray arrow for Mac) to expand the information layers in the side pane, then complete the following table of information.

Date, time:	
Azimuth, altitude:	
In constellation:	
Bayer designation (alpha, beta, gamma, etc.):	
Distance from Sun/observer (give units):	
Radius:	
Temperature:	
Apparent magnitude:	
Absolute magnitude:	
Luminosity:	

Provide a sketch of the constellation in which this star is located, indicating the exact location of the star.

PART 4: FINDING YOUR OWN EXAMPLES

In this section, we are interested in finding our own examples of typical stars. Zoom all the way out, return to just after sunset by following the instructions given in Part 2, and scan your current evening sky for some characteristic stars.* You may do this either by using the direction buttons N, S, E, W, and Z on the **button bar**, or by simply "grabbing" the sky with the **hand tool** and dragging while holding the left mouse button down. You may need to zoom in and out a bit to see the dimmer stars. You may also advance time forward to later in the evening if you wish.

- If you wish to turn other objects "off," which may distract from our survey of stars, click the plus sign (gray arrow for Mac) to expand the **solar system** layer of the VIEW OPTIONS **side pane** and turn off items such as **satellites**, **planets-moons**, **comets**, **asteroids**, etc. You can increase the number of stars being labeled on your screen by Starry Night by expanding the **stars** layer of the VIEW OPTIONS **side pane**, hovering your pointer over the word **stars**, and then clicking to change viewing options. Slide the **labels** bar all the way to the right, then click **OK**. Don't slide the other bars to increase the number of stars shown and/or labeled (at least not yet). Starry Night has such a large database that doing this can dramatically decrease your system's performance.

Look for bright stars, then right click (control click for Mac), being careful to hover the mouse exactly over the object as best you can, and select **show info**. Detailed information about this star will now be displayed in the side pane. Expand all layers by clicking the small plus signs to the left (gray arrows for Mac).

STAR #1: Use the information provided by Starry Night to complete the following table:
(Try not to select any of the same examples already given in this activity.)

Date, time:	
Azimuth, altitude:	
Name and catalog number:	
In constellation:	
Bayer designation (alpha, beta, gamma, etc.):	
Distance from the Sun/observer (give units):	
Radius:	
Temperature:	
Apparent magnitude:	
Absolute magnitude:	
Luminosity:	

Provide a sketch of the constellation in which this star is located, indicating the stars exact location.

<u>STAR #2</u>: Use the information provided by Starry Night to complete the following table:
(Try not to select any of the same examples already given in this activity.)

Date, time:	
Azimuth, altitude:	
Name and catalogue number:	
In constellation:	
Bayer designation (alpha, beta, gamma, etc.):	
Distance from the Sun/observer (give units):	
Radius:	
Temperature:	
Apparent magnitude:	
Absolute magnitude:	
Luminosity:	

Provide a sketch of the constellation in which this star is located, indicating its exact location.

<u>STAR #3</u>: Use the information provided by Starry Night™ College to complete the following table:
(Try not to select any of the same examples already given in this activity.)

Date, time:	
Azimuth, altitude:	
Name and catalog number:	
In constellation:	
Bayer designation (alpha, beta, gamma, etc.):	
Distance from Sun/observer (give units):	
Radius:	
Temperature:	
Apparent magnitude:	
Absolute magnitude:	
Luminosity:	

Provide a sketch of the constellation in which this star is located, indicating its exact location.

PART 5: [OPTIONAL] LUMINOSITY CALCULATIONS

What does the magnitude and luminosity information you found for each of these stars mean? The primary factors that affect a star's brightness are distance, temperature, and size.

Brightness varies by the inverse square of distance; thus a star twice as far away is one-quarter $(1/2^2)$ the brightness. A star three times as far away is one-ninth $(1/3^2)$ the brightness.

Brightness varies by the fourth power of temperature where hot stars are brighter than dim stars. A star twice as hot is 16 (2^4) times brighter. A star three times as hot is 81 (3^4) times brighter.

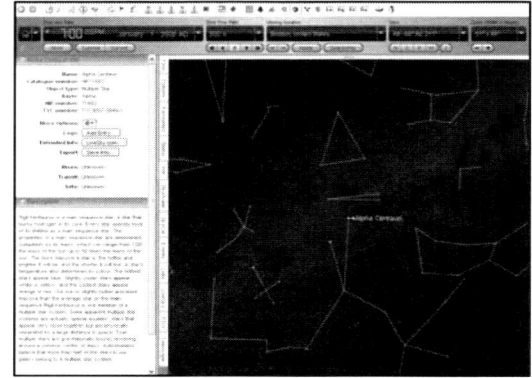

Brightness also varies by the square of the radius. Although the star is a sphere, only the cross-sectional solid angle contributes to the brightness we see. Thus a star twice as big is four (2^2) times as bright. A star three times as big is nine (3^2) times as bright.

Use the relationships above to determine how much brighter (or dimmer) a hypothetical star with the given characteristics would be as compared to Alpha Centauri, a Sun-like star a little over 4 light-years away. These are "order of magnitude" calculations (to keep the math simple). For simplicity, state whether the configuration described for the star at the left is *brighter* or *dimmer* than for Alpha Centauri. Alternatively, you may express your answer numerically.

Characteristics of Hypothetical Star *(to be compared to Alpha Centauri)*	Multiple of Alpha Centauri's Brightness *(expressed qualitatively or quantitatively)*
2 × Distance, 1 × Temperature, 2 × Size	
2 × Distance, 2 × Temperature, 2 × Size	
100 × Distance, 3 × Temperature, 10 × Size (Blue Giant Star)	
100 × Distance, 4 × Temperature, 10 × Size (Blue Supergiant Star)	
100 × Distance, (½)× Temperature, 100 × Size (Red Supergiant Star)	
100 × Distance, 10 × Temperature, (1/100) × Size (White Dwarf Star)	

PART 6: [OPTIONAL] LOTS AND LOTS OF STARS

To gain an appreciation for how many stars are out there, do the following.

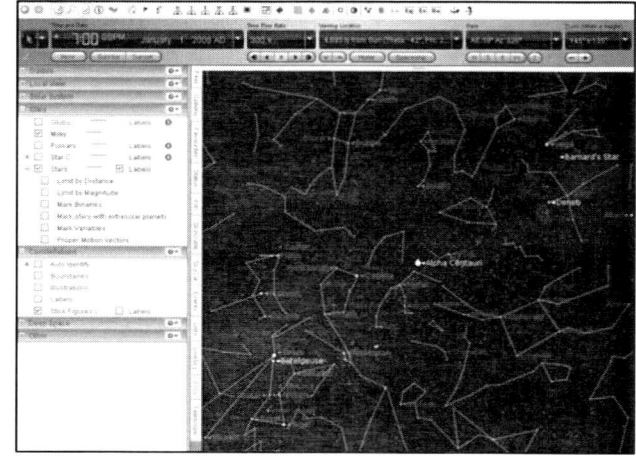

Depending on your system's capabilities, you can increase the number of stars displayed. This can give a real appreciation for the vastness of space and the number and variety of star systems out there. Binoculars, telescopes, and planetarium programs such as this can help you to explore the mysteries of the sky.

You can increase the number of stars shown on your screen by expanding the **stars** layer of the OPTIONS **side pane**, hovering your pointer over the word **stars**, and then clicking to change viewing options. Slide the **number of stars** slider all the way to the right, then click **OK**.

If your system is moving significantly slow, be patient, then click the **back space** button at the upper left of the **button bar** to restore your settings to what they were previously.

There are some 200 billion stars in our Milky Way galaxy, only a small fraction of which can be seen by the world's most powerful telescopes, and an even smaller fraction by the naked eye. For illustrative purposes, let's say you could see and count all the stars in our galaxy. If you counted one star a second, nonstop, 24 hours a day (no breaks, no mistakes, continuously), how long would it take you to count all 200 billion stars? Express your answer in years. How long would it take to count the 300 billion stars of our nearest neighbor, the Andromeda galaxy?

Years needed to count stars in Milky Way (one a second):	
Years needed to count stars in Andromeda galaxy (one a second):	

Are you surprised? These galaxies are not especially large; and keep in mind, there are billions and billions of galaxies.

Activity 20—VARIABLE STARS

*These activities are designed to work with the Starry Night software that comes with your text, from any home location you choose, and with the current date and time, unless indicated otherwise. You may always revert to factory default settings by clicking **FILE/ preferences**, then selecting **factory defaults** as needed. You may also undo a command or series of commands on the PC by clicking the **back** button at the top left of the **button bar**. You should refer to the key given at the beginning of this booklet for clarification of "on screen" buttons, controls, and functions. PC **button bar** items can all be accessed through the **menu**. "Right click" on the PC is equivalent to "control click" on the Mac. All activities assume that OpenGL graphics capabilities are enabled on your computer.*

PART 1: DIFFERENT TYPES OF STARS

The stars you see at night are all very different. There are stars (stars by themselves); variable stars (stars which vary in brightness); binary stars (pairs of stars); multiple star systems (all gravitationally bound together); main sequence stars; red and blue giants; red and blue supergiants; and a host of other star-like objects, such as brown dwarfs, white dwarfs, neutron stars, pulsars, and black holes. With so much variety in the sky, how do we know what we are looking at?

This activity will focus on variable stars. Variable stars change their brightness, most often in a predictable time period that relates to the age and physical properties of the star. Variable stars are important to astronomers in that they allow for theoretical models to predict a star's mass, age, and brightness. There are different reasons why a star may be variable. The star may be unstable during formation, may be unstable during certain nuclear fusion stages while nearing the end of its interstellar life, and/or may simply be eclipsing another companion star as part of its orbit in an eclipsing binary system.

PART 2: VARIABLE STARS OF OUR CURRENT EVENING SKY

You should begin this activity at sunset. An easy way to do this is to click the drop-down menu to the right of the **date & time** field on the **control panel**, and select **sunset**. Look toward the west by clicking the **W viewing direction** button located on the **button bar** across the top of your screen, or by simply keying in the letter W (Mac users should refer to the button bar commands given at the beginning of this booklet). The screen will pan toward the west. Select a playing speed of **300×** normal time by clicking the drop-down menu at the right of the **time speed** field. Click the STOP **time mode** button when the Sun has set, the stars have come out, and dusk is almost over. Then click on the **constellations** button to show the constellations, and click the **labels** button to label stars and objects.

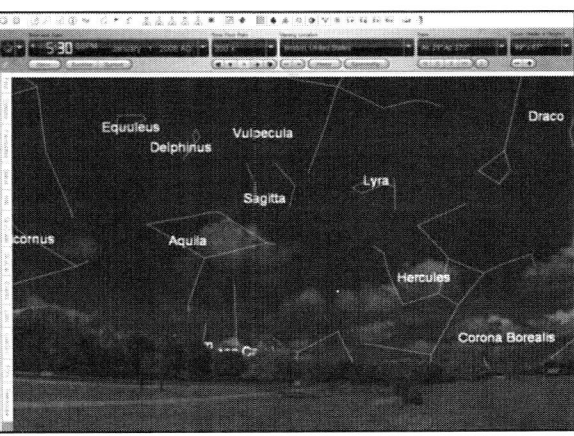

Click the OPTIONS tab on the left **side pane** and expand the **stars** layer by clicking on the small plus sign to the left (gray arrow for Mac). Click the **mark variables** check box. You should now see many variables identified on your screen. Since the marker for variables is subtle, you may want to toggle the **mark variables** check box on and off while closely watching the screen.

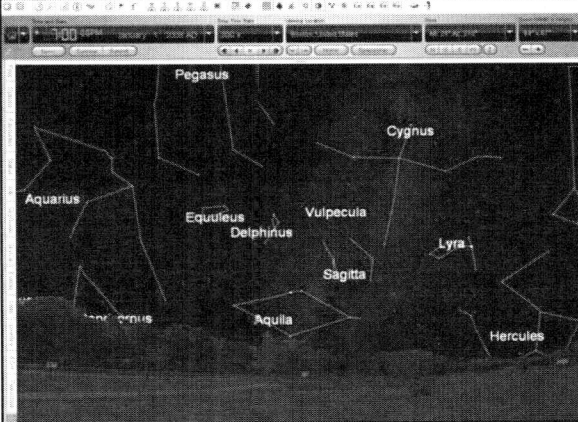

[OPTIONAL] You can increase the number of stars being labeled on your screen by Starry Night by expanding the **star** layer of the OPTIONS **side pane**, then hover your pointer over the word **stars**, then click to change viewing options. Slide the **labels** bar all the way to the right, then click **OK**. Don't slide the other bars to increase the number of stars shown and/or labeled. Starry Night has such a large database that doing this can dramatically decrease your system's performance.

[OPTIONAL] If you wish to turn "off" other objects that may distract from our survey of variable stars, click the plus sign (gray arrow for Mac) to expand the **solar system** layer of the OPTIONS **side pane** and turn off items such as **satellites**, **planets-moons**, **comets**, **asteroids**, etc.

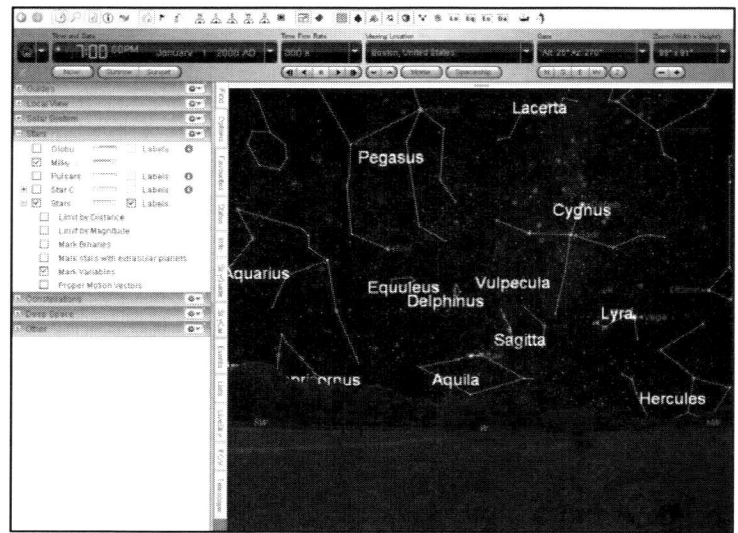

PART 3: EXAMPLES OF VARIABLES

Let's look at some well-known examples. Astronomers always have a few favorites. Which of these is your favorite?

<u>ALGOL the DEMON STAR (HIP14576):</u>

Algol is the famous "Demon Star" of mythology. Perseus is said to have cut off the head of Medusa. As Perseus flies across the heavens with Mercury's winged sandals, the gruesome head of Medusa is said to still glow, changing its brightness by a few orders of magnitude.

Click the FIND tab on the left **side pane**. In the search field at the top, type in ALGOL. Right click (control click for Mac) on the object and select **show info**, click the plus signs (gray arrow for Mac) to expand the information layers in the side pane, then complete the following table of information.

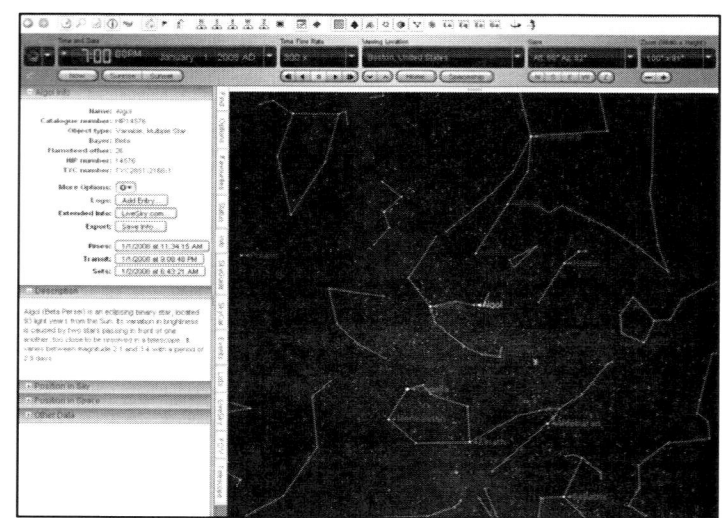

Complete the table below for Algol, the Demon Star:

Date, time:	
Azimuth, altitude:	
In constellation:	
Distance from observer (give units):	
Bayer designation (alpha, beta, gamma, etc):	
Apparent magnitude:	
Absolute magnitude:	
Variability:	
Radius:	
Temperature:	
Luminosity:	

Provide a sketch of the constellation that shows the location of this star:

| |
| |

POLARIS the NORTH STAR (HIP11767):

Polaris, our North Star, is no ordinary star. It is both a variable and double/multiple star. Many stars are variable and double, but Polaris is significant because of its position, close to a degree from the North Celestial Pole.

Click the FIND tab on the left **side pane**. In the search field at the top, type in POLARIS. Right click (control click for Mac) on the object and select **show info**, click the plus signs (gray arrow for Mac) to expand the information layers in the side pane, then complete the following table of information.

Complete the table below for Polaris, the North Star:

Date, time:	
Azimuth, altitude:	
In constellation:	
Distance from observer (give units):	
Bayer designation (alpha, beta, gamma, etc.):	
Apparent magnitude:	
Absolute magnitude:	
Variability:	
Radius:	
Temperature:	
Luminosity:	

Provide a sketch of the constellation that shows the location of this star:

| |
| |

PART 4: FINDING YOUR OWN EXAMPLES

In this section, we are interested in finding our own examples of variable stars. Zoom all the way out, return to just after sunset by following the instructions given in Part 2, and scan your current evening sky for some characteristic variables. You may do this either by using the direction buttons N, S, E, W, and Z on the **button bar**, or by simply "grabbing" the sky with the **hand tool** and dragging while holding the left mouse button down. You may need to zoom in and out a bit to see the dimmer objects. You may also advance time forward to later in the evening if you wish.

Look for variable Stars and note the star's variability by reading what the **variability** field states while hovering your mouse over it. Right click (control click for Mac) on this object (careful to hover the mouse exactly over the object as best you can) and select **show info**. Detailed information about this star will now be displayed in the side pane. Click the plus sign (gray arrow for Mac) to expand the information layers in the side pane.

STAR #1: Use the information provided by Starry Night to complete the following table:
(Try not to select any of the same examples already given in this activity.)

Date, time:	
Azimuth, altitude:	
Name and catalog number:	
In constellation:	
Distance from observer (give units):	
Bayer designation (alpha, beta, gamma, etc.):	
Apparent magnitude:	
Absolute magnitude:	
Variability:	
Radius:	
Temperature:	
Luminosity:	

Provide a sketch of the constellation that shows where this star is located:

<u>STAR #2</u>: Use the information provided by Starry Night to complete the following table:
(Try not to select any of the same examples already given in this activity.)

Date, Time:	
Azimuth, altitude:	
Name and catalog number:	
In constellation:	
Distance from observer (give units):	
Bayer designation (alpha, beta, gamma, etc.):	
Apparent magnitude:	
Absolute magnitude:	
Variability:	
Radius:	
Temperature:	
Luminosity:	

Provide a sketch of the constellation that shows where this star is located:

<u>STAR #3</u>: Use the information provided by Starry Night to complete the following table:
(Try not to select any of the same examples already given in this activity.)

Date, time:	
Azimuth, altitude:	
Name and catalogue number:	
In constellation:	
Distance from observer (give units):	
Bayer designation (alpha, beta, gamma, etc.):	
Apparent magnitude:	
Absolute magnitude:	
Variability:	
Radius:	
Temperature:	
Luminosity:	

Provide a sketch of the constellation showing where this star is located:

PART 4: [OPTIONAL] LUMINOSITY DISCUSSION AND CALCULATIONS

What does all this information you found for each of these variables mean? Stars may vary in brightness for various reasons. Some may be pulsating variables that change in size, pulsating due to instabilities resulting from nuclear fusion reactions that often occur later in their stellar lives. Others may be eclipsing binaries, where an often smaller, less bright companion passes in front (or behind) the star, changing its brightness as perceived by us far away. Keep in mind that in this case, the stars themselves may not vary in brightness, but rather it's the effect of the stellar eclipse that influences what we see. Finally, stars may vary in brightness due to violent outbursts or activity, such as collisions, novae, or supernovae. Pulsating variables and eclipsing binaries change their brightness in predictable cycles with a specific variability period. Stars that go nova or supernova produce a single explosion and/or a quick series of outbursts, then may fade away over many years.

The primary factors that influence a star's brightness (and thus its variability) are distance, temperature, and size.

Brightness varies by the inverse square of distance, thus a star twice as far away is one-quarter $(1/2^2)$ the brightness. A star three times as far away is one-ninth $(1/3^2)$ the brightness.

Brightness varies by the fourth power of temperature where hot stars are brighter than dim stars. A star twice as hot is 16 (2^4) times brighter. A star three times as hot is 81 (3^4) times brighter.

Brightness varies by the square of the radius. Although the star is a sphere, only the cross-sectional solid angle contributes to the brightness we see. Thus a star twice as big is four (2^2) times as bright. A star three times as big is nine (3^2) times as bright.

Use the relationships above to determine how much brighter (or dimmer) a star with the given characteristics would be as compared to Alpha Centauri, a Sun-like star a little over four light-years away. These are "order of magnitude" calculations (to keep the math simple).

$2 \times$ Distance, $1 \times$ Temperature, $2 \times$ Size	
$2 \times$ Distance, $2 \times$ Temperature, $2 \times$ Size	
$100 \times$ Distance, $3 \times$ Temperature, $10 \times$ Size (Blue Giant Star)	
$100 \times$ Distance, $4 \times$ Temperature, $10 \times$ Size (Blue Supergiant Star)	
$100 \times$ Distance, $(\frac{1}{2}) \times$ Temperature, $100 \times$ Size (Red Supergiant Star)	
$100 \times$ Distance, $10 \times$ Temperature, $(1/100) \times$ Size (White Dwarf Star)	

Activity 21—BINARY STARS

*These activities are designed to work with the Starry Night software that comes with your text, from any home location you choose, and with the current date and time, unless indicated otherwise. You may always revert to factory default settings by clicking **FILE/ preferences**, then selecting **factory defaults** as needed. You may also undo a command or series of commands on the PC by clicking the **back** button at the top left of the **button bar**. You should refer to the key given at the beginning of this booklet for clarification of "on screen" buttons, controls, and functions. PC **button bar** items can all be accessed through the **menu**. "Right click" on the PC is equivalent to "control click" on the Mac. All activities assume that OpenGL graphics capabilities are enabled on your computer.*

PART 1: DIFFERENT TYPES OF STARS

The stars you see at night are all very different. There are stars (stars by themselves); variable stars (stars which vary in brightness); binary stars (pairs of stars); multiple star systems (all gravitationally bound together); main sequence stars; red and blue giants; red and blue supergiants; and a host of other star-like objects such as brown dwarfs, white dwarfs, neutron stars, pulsars, and black holes. With so much variety in the sky, how do we know what we are looking at?

This activity will focus on binary stars and multiple star systems. More than half the stars in our evening sky are binary or multiple star systems. They often appear as a single star due to their great distance from us. Some can be resolved through telescopes, others require sophisticated equipment and data analysis to detect their presence.

Binary stars are important to astronomers in that they allow for theoretical models to predict a star's mass and orbital characteristics. There are four different types of binaries.

Optical binaries are stars that appear to be next to each other, but are in fact far apart radially and thus not gravitationally bound. There is no way of knowing visually if the two stars, which appear close together, are the same distance away. One may be close while the other is far. Special equipment and data analysis is required to determine each star's distance, and only then can we ascertain if they are gravitationally bound together into a binary star system.

True (visual) binaries are stars that are visually next to each other in the sky, and also about the same distance away, indicating that they are gravitationally bound together into a binary star system.

Spectroscopic binaries are binary systems that cannot be resolved optically with even the most powerful telescopes, but are determined to be binaries through spectroscopic measurements. Spectroscopes allow for the light to be split up into its constituent colors. The colors we see indicate much about a star's physical properties. Careful measurements of the wavelengths of spectral lines can tell us the star's radial velocity (the speed with which it is approaching or receding from us). Two-star binary systems will produce two sets of spectroscopic measurements or lines (one approaching, the other receding) from the light of what appears to be a single star. Thus, we may deduce that what appears as a single star in even the most powerful telescopes, is in fact a binary star system whose companions are simply too close together, and the system too far away, to be resolved visually.

Eclipsing binaries are binary systems oriented such that one member of the system will eclipse the other member, producing a variable light curve of a characteristic shape and periodicity. Such binaries may not be discernible visually and are identified by analysis of the light curve produced.

PART 2: BINARY STARS OF OUR CURRENT EVENING SKY

You should begin this activity at sunset. An easy way to do this is to click the drop-down menu to the right of the **date & time** field on the **control panel**, and select **sunset**. Look toward the west by clicking the **W viewing direction** button located on the **button bar** across the top of your screen, or by simply keying in the letter W (Mac users should refer to the button bar commands given at the beginning of this booklet). The screen will pan toward the west. Select a playing speed of **300×** normal time by clicking the drop-down menu at the right of the **time speed** field. Click the stop **time mode** button when the Sun has set, the stars have come out, and dusk is almost over. Then click on the **constellations** button to show the constellations, and click the **labels** button to label stars and objects.

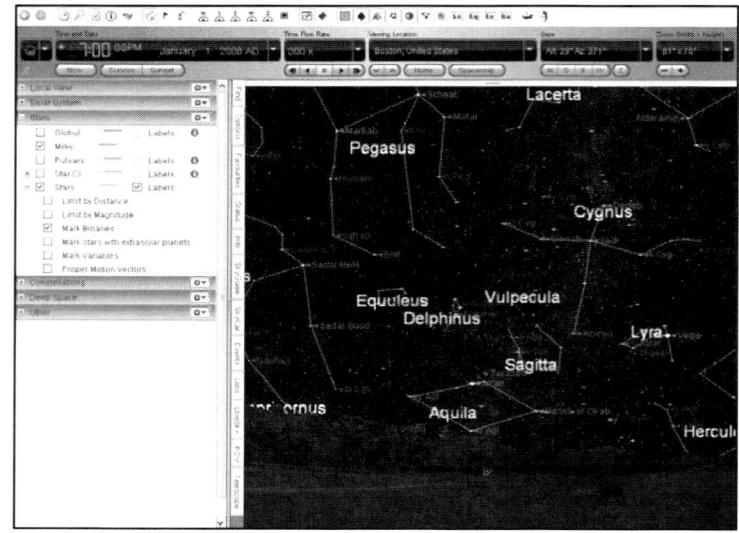

Click the OPTIONS tab on the left **side pane** and expand the **stars** layer by clicking on the small plus sign to the left (gray arrow for Mac). Click the **mark binaries** check box.

You should now see many binaries identified on your screen. Since the marker for binaries is subtle, you may want to toggle the **mark binaries** check box on and off while closely watching the screen.

[OPTIONAL] You can increase the number of binaries being labeled on your screen by Starry Night by clicking the plus sign (gray arrow for Mac) to expand the **stars** layer of the OPTIONS **side pane**, then hover your pointer over the word **stars**, then click to change viewing options. Slide the **labels** bar all the way to the right, then click **OK**. Don't slide the other bars to increase the number of stars shown and/or labeled (at least not yet). Starry Night™ Pro has such a large database that doing this can dramatically decrease your system's performance.

[OPTIONAL] If you wish to turn "off" other objects that may distract from our survey of binary stars, click the plus sign (gray arrow for Mac) to expand the **solar system** layer of the OPTIONS **side pane** and turn off items such as **satellites**, **planets-moons**, **comets**, **asteroids**, etc.

PART 3: EXAMPLES OF BINARY STARS

Let's look at some well-known examples. Astronomers always have a few favorites. Which of these is your favorite?

<u>MIZAR & ALCOR (HIP65378 / STF 1744 & HIP65477):</u>

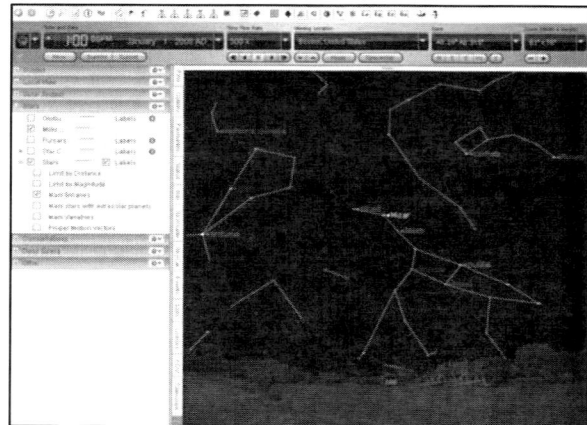

Click the FIND tab on the left **side pane**. In the search field at the top, type in MIZAR. Right click (control click for Mac) on the object and select **show info**, click the plus signs (gray arrow for Mac) to expand the information layers in the side pane, then complete the following table of information.

When you zoom in (use the **zoom controls** at the right of the **control panel**), you will see three stars close together as shown here. In this image, Mizar is at the lower part of the screen, Alcor at the upper part of the screen, and a third star in between and to the left a bit. Mizar and Alcor are relatively close together. You can click, hold, and drag the pointer from one star to the other to measure the angular separation between Mizar and Alcor. You will find that Alcor is about three light-years from Mizar. The third star has an unknown distance and is not thought to be gravitationally related to either Mizar or Alcor. Together, in a telescope's low-power eyepiece, we see all three stars; however, Mizar and Alcor are themselves both binary stars, with Mizar and its

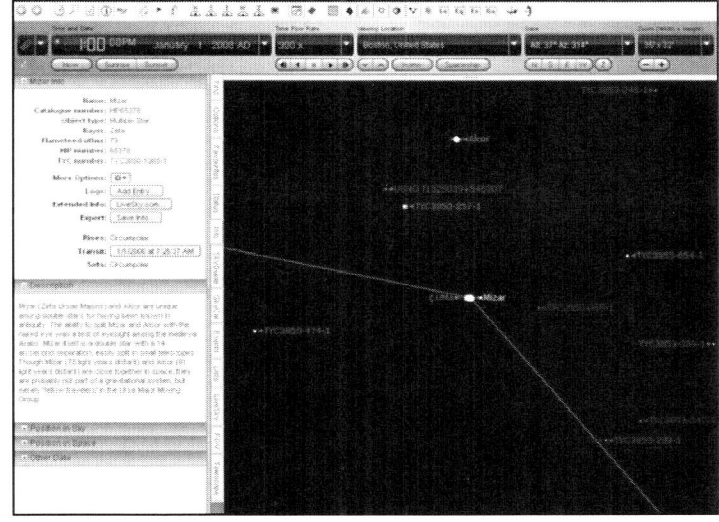

companion being spectroscopic binaries as well. So, we see what appears to be three stars together in the telescope field of view, two of them being binaries, with Mizar being a multiple star system of four stars. It's interesting to note that Mizar and Alcor are visible to the naked eye. Ancient cultures may have used Mizar and Alcor as a way to test their eyesight. Try this out yourself. Can you see both Mizar and Alcor on a dark night? Check with your friends to see who can make this combo out. Then try it with binoculars and/or telescopes.

Complete the table below, but only for Mizar and its companion (don't worry about Alcor):

Date, time:	
Azimuth, altitude:	
In constellation:	
Distance from observer (give units):	
Bayer designation (alpha, beta, gamma, etc.):	
Apparent magnitude of primary:	
Absolute magnitude of primary:	
Luminosity of primary:	
Angular separation (as given in info layer):	
Angular separation (measured by you):	

Provide a sketch of the constellation that shows where this binary is located:
(Also show the relative position of the binary as you saw it when zooming in.)

POLARIS the NORTH STAR (STF 93):

Polaris, our North Star, is no ordinary star. It is both a variable and double/multiple star. Many stars are variable and double, but Polaris is significant because of its position (within a degree of the *North Celestial Pole).*

Click the FIND tab on the left **side pane**. In the search field at the top, type in POLARIS. Right click (control click for Mac) on the object and select **show info**, click the plus signs (gray arrow for Mac) to expand the information layers in the side pane, then complete the following table of information.

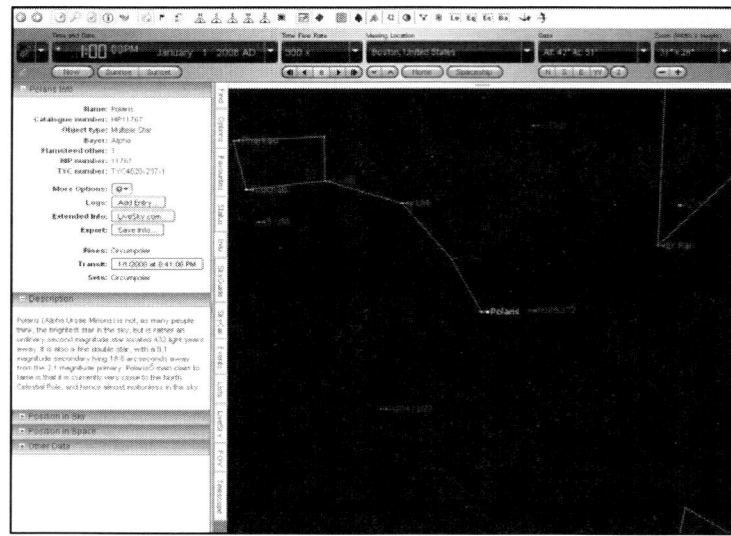

Complete the table below for Polaris and its companion(s):

Date, time:	
Azimuth, altitude:	
In constellation:	
Distance from observer (give units):	
Bayer designation (alpha, beta, gamma, etc.):	
Apparent magnitude of primary:	
Absolute magnitude of primary:	
Luminosity of primary:	
Angular separation (as given in info layer):	
Angular separation (measured by you):	

Provide a sketch of the constellation that shows where this binary is located:
(Also show the relative position of the binary as you saw it when zooming in.)

ALBIREO (STF 43):

Click the FIND tab on the left **side pane**. In the search field at the top, type in ALBIREO. Right click (control click for Mac) on the object and select **show info**, click the plus signs (gray arrow for Mac) to expand the information layers in the side pane, then complete the following table of information.

When you zoom in (use the **zoom controls** at the right of the **control panel**), you will see two stars close together as shown here. You should note that Albireo and its companion are both binaries as well. Albireo appears slightly reddish, and its companion appears somewhat bluish. This color contrast makes for a strikingly beautiful telescopic sight.

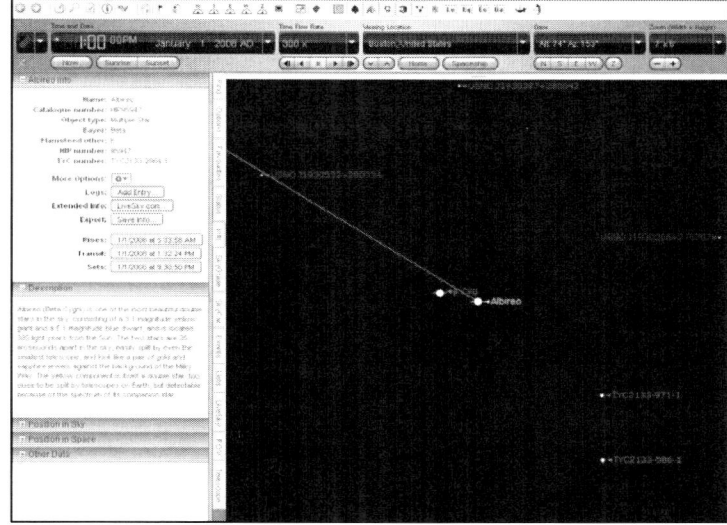

Complete the table below for Albireo and its companion:

Date, time:	
Azimuth, altitude:	
In constellation:	
Distance from observer (give units):	
Bayer designation (alpha, beta, gamma, etc.):	
Apparent magnitude of primary:	
Absolute magnitude of primary:	
Luminosity of primary:	
Angular separation (as given in info layer):	
Angular separation (measured by you):	

Provide a sketch of the constellation showing where this binary is located:
(Also show the relative position of the binary as you saw it when zooming in.)

[OPTIONAL] COLOR COMPARISON between ALBIREO and its COMPANION:

Compare the *radius*, *temperature*, and *B-V color* for Albireo and its companion. You will see that Albireo is a red giant or supergiant star (see how many times larger than our Sun it is). Also, note that Albireo's companion is a hot blue star with a very high surface temperature and negative *B-V color* index. The *B-V color* index will be positive for reddish stars, and negative for bluish stars.

Complete the table below for Albireo and its companion:

Star name or designation:	Albireo	Companion
Radius:		
Temperature:		
B-V color:		

PART 4: FINDING YOUR OWN EXAMPLES

In this section, we are interested in finding our own examples of binary star systems. Zoom all the way out, return to just after sunset following the instructions given in Part 2, and scan your current evening sky for some characteristic binaries.* You may do this either by using the **viewing direction** buttons N, S, E, W, and Z on the **button bar**, or by simply "grabbing" the sky with the **hand tool** and dragging while holding the left mouse button down. You may need to zoom in and out a bit to see the dimmer objects. You may also advance time forward to later in the evening if you wish.

*If you wish to turn other objects "off," which may distract from our survey of binary stars, click the plus sign (gray arrow for Mac) to expand the **solar system** layer of the OPTIONS **side pane** and turn off items such as **satellites**, **planets-moons**, **comets**, **asteroids**, etc.

Look for binary stars and note the star's *angular separation* by reading what the **double/multiple** field states while hovering your mouse over it. Right click (control click for Mac) on the object (careful to hover the mouse exactly over the object as best you can) and select **show info**. Detailed information about this star will now be displayed in the side pane. Click the plus signs (gray arrow for Mac) to expand the information layers in the side pane.

Use the **zoom control** at the right of the **control panel** to zoom in on the binary star system and measure the *angular separation* manually. Not all binaries can be visually resolved, so you won't be able to do this for all binaries you find. Hover your mouse pointer over the star until you see an arrow appear (watch as the hand turns to an arrow), then click with the left mouse button and drag away from the star. A line will appear and information indicating the angular separation will be shown. When you drag the line over to another star, in some cases for stars whose 3-D data coordinates are known to Starry Night, the distance will be displayed as well. Use this technique to find the angular separation (and distance if displayed) of the binaries you view.

STAR #1: Use the information provided by Starry Night to complete the following table:
(Try not to select any of the same examples already given in this activity.)

Date, time:	
Azimuth, altitude:	
Name and catalog number:	
In constellation:	
Distance from observer (give units):	
Bayer designation (alpha, beta, gamma, etc.):	
Apparent magnitude of primary:	
Absolute magnitude of primary:	
Luminosity of primary:	
Angular separation (as given in info layer):	
Angular separation (measured by you):	

Provide a sketch of the constellation showing where this binary is located:
(Also show the relative position of the binary as you saw it when zooming in.)

STAR #2: Use the information provided by Starry Night to complete the following table:
(Try not to select any of the same examples already given in this activity.)

Date, time:	
Azimuth, altitude:	
Name and catalog number:	
In constellation:	
Distance from observer (give units):	
Bayer designation (alpha, beta, gamma, etc.):	
Apparent magnitude of primary:	
Absolute magnitude of primary:	
Luminosity of primary:	
Angular separation (as given in info layer):	
Angular separation (measured by you):	

Provide a sketch of the constellation showing where this binary is located:
(Also show the relative position of the binary as you saw it when zooming in.)

STAR #3: Use the information provided by Starry Night to complete the following table:
(Try not to select any of the same examples already given in this activity.)

Date, time:	
Azimuth, altitude:	
Name and catalog number:	
In constellation:	
Distance from observer (give units):	
Bayer designation (alpha, beta, gamma, etc.):	
Apparent magnitude of primary:	
Absolute magnitude of primary:	
Luminosity of primary:	
Angular separation (as given in info layer):	
Angular separation (measured by you):	

Provide a sketch of the constellation showing where this binary is located:

(Also show the relative position of the binary as you saw it when zooming in.)

Activity 22—OPEN CLUSTERS

These activities are designed to work with the Starry Night software that comes with your text, from any home location you choose, and with the current date and time, unless indicated otherwise. You may always revert to factory default settings by clicking FILE/ preferences, then selecting factory defaults as needed. You may also undo a command or series of commands on the PC by clicking the back button at the top left of the button bar. You should refer to the key given at the beginning of this booklet for clarification of "on screen" buttons, controls, and functions. PC button bar items can all be accessed through the menu. "Right click" on the PC is equivalent to "control click" on the Mac. All activities assume that OpenGL graphics capabilities are enabled on your computer.

PART 1: OPEN CLUSTERS

This activity will focus on open clusters. Open clusters are groups of stars that have recently formed from the same interstellar cloud of gas and dust. They are relatively young in age and some can be seen with the naked eye and/or with binoculars.

PART 2: FINDING OPEN CLUSTERS IN OUR NIGHT SKY

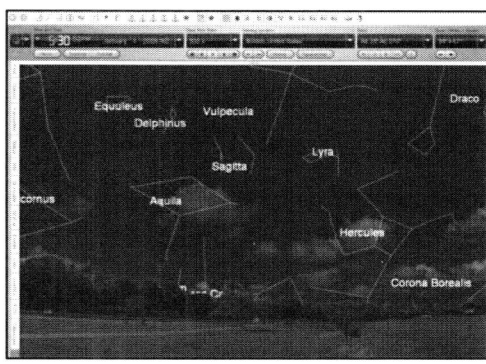

You should begin this activity at sunset. An easy way to do this is to click the drop-down menu to the right of the **date & time** field on the **control panel**, and select **sunset**. Look toward the west by clicking the **W viewing direction** button located on the **button bar** across the top of your screen, or by simply keying in the letter W (Mac users should refer to the button bar commands given at the beginning of this booklet). The screen will pan toward the west. Select a playing speed of **300×** normal time by clicking the drop-down menu at the right of the **time speed** field. Click the STOP **time mode** button when the Sun has set, the stars have come out, and dusk is almost over. Then click on the **constellations** button to show the constellations.

Click the OPTIONS tab on the left **side pane** and expand the **deep space** layer by clicking on the small plus sign to the left (gray arrow for Mac). Activate the labels for both the **bright NGC objects** and **Messier objects** by clicking the **labels** box to the right. Then click the information icon **(i)** to the far right to display descriptive information on these Starry Night databases.

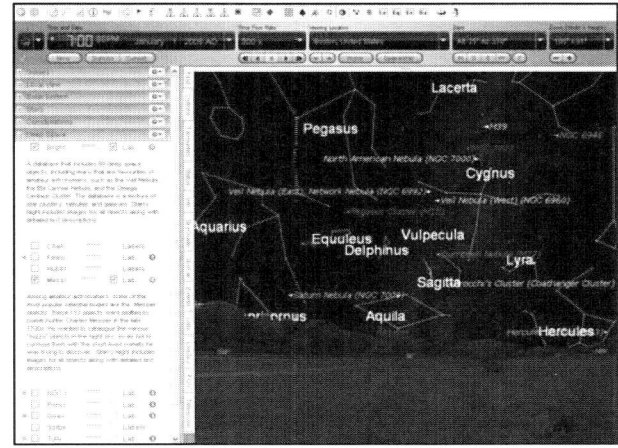

You should now see numerous deep-space objects identified on your screen. These are some of the most famous telescope viewing objects in the sky.

[OPTIONAL] For a more complete listing of open clusters, you may turn on and label additional open clusters by clicking the box to the left of the **NGC-IC database** listing in the **other** layer of the OPTIONS **side pane**. Don't click the **label** box to the right. Click the plus sign to the left (gray arrow for Mac), then select **open cluster** by checking the box to the left while unchecking all the others. Again, don't click the **label** box to the right. A small marker will now identify additional open clusters on your screen. For this activity, scan the sky only for those that are also labeled, since only these will display an image in Starry Night. Take careful note of the fact that this is an overlapping database. This means that some of the objects in this database are already displayed in Starry Night by other databases. This can lead to confusion since the coordinates may not be exactly the same, thus appearing to indicate that the same object is in two slightly different places at the same time. Although not exact, the additional database markers can help us locate objects. As you zoom in, you should ignore the marker and look in the general vicinity for the intended object. Once you locate it, right click (control click for Mac) and center it on your screen and continue zooming in. Be careful not to confuse open clusters with other Messier and NGC objects. The on screen info will confirm the **object type** when you hover your mouse over it.

PART 3: EXAMPLES OF OPEN CLUSTERS

Let's look at some well-known examples. Astronomers always have a few favorites. Which of these is your favorite?

THE PLEIADES OPEN CLUSTER (M45):

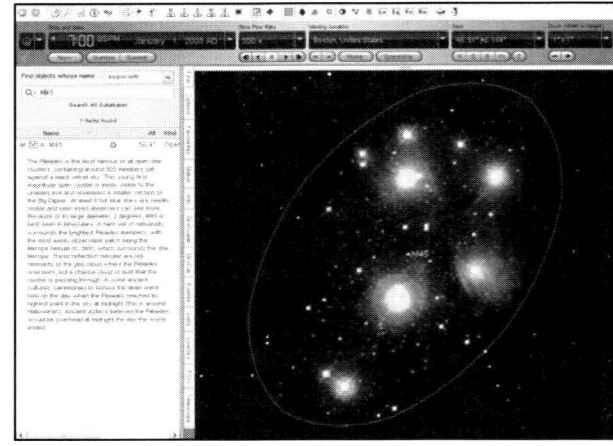

Click the FIND tab on the left **side pane**. In the search field at the top, type in M45. More than one listing may appear. Click the information icon (**i**) to the right for all entries, then grab the right side double line and click and drag to the right to expand the text box as far as it will go. You will note that the listings are from different databases. This simply means that Starry Night has this object appearing in multiple (different) databases (or lists). All refer to the same object; however, the information given may be slightly different and the coordinates may vary slightly as well. Double-click on the listing associated with the **Messier object** database. Zoom in on the object using the **zoom controls** at the right of the **control panel**. Right click (control click for Mac) on the object and select **show info**, click the plus sign (gray arrow for Mac) to expand all sections, then complete the following table of information.

Date, time:	
Azimuth, altitude:	
Distance from observer (don't forget units):	
Apparent magnitude:	
Angular size (give units—degrees, arc minutes, etc.):	

What do you find most interesting? Indicate some additional info about this object:

THE PERSEUS DOUBLE CLUSTER (NGC 884):

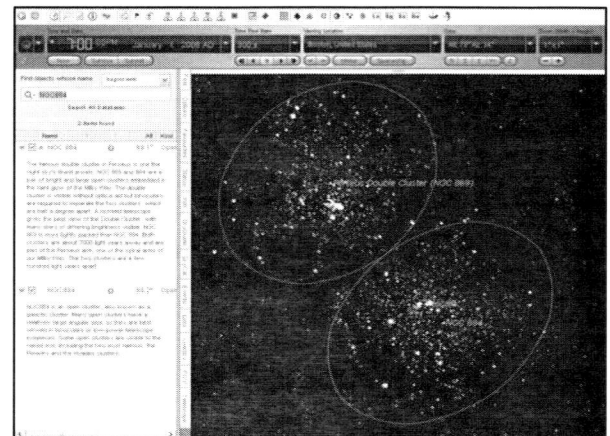

Click the FIND tab on the left **side pane**. In the search field at the top, type in NGC 884. More than one listing may appear. Click the information icon (**i**) to the right for all entries, then grab the right side double line and click and drag to the right to expand the text box as far as it will go. You will note that the listings are from different databases. Starry Night has this object appearing in multiple (different) databases (or lists). All refer to the same object; however, the information given may be slightly different and the coordinates may vary slightly as well. Double click on the listing associated with the **bright NGC objects** database. Adjust the **zoom controls** as needed. Right click (control click for Mac) on the object and select **show info**, click the plus sign (gray arrow for Mac) to expand all sections, then complete the following table of information.

Date, time:	
Azimuth, altitude:	
Distance from observer (don't forget units):	
Apparent magnitude:	
Angular size (give units—degrees, arc minutes, etc.):	

What do you find most interesting? Indicate some additional info about this object:

THE WILD DUCK CLUSTER (M11):

Click the FIND tab on the left **side pane**. In the search field at the top, type in M11. More than one listing may appear. Click the information icon **(i)** to the right for all entries, then grab the right side double line and click and drag to the right to expand the text box as far as it will go. You will note that the listings are from different databases. Starry Night has this object appearing in multiple (different) databases (or lists). All refer to the same object; however, the information given may be slightly different and the coordinates may vary slightly as well. Double-click on the listing associated with the **Messier objects** database. Adjust the **zoom controls** as needed. Right click (control click for Mac) on the object and select **show info**, click the plus sign (gray arrow for Mac) to expand all sections, then complete the following table of information.

Date, time:	
Azimuth, altitude:	
Distance from observer (don't forget units):	
Apparent magnitude:	
Angular size (give units—degrees, arc minutes, etc.):	

What do you find most interesting? Indicate some additional info about this object:

PART 4: FINDING YOUR OWN EXAMPLES

In this section, we are interested in finding our own examples of open clusters. Zoom all the way out, return to just after sunset by following the instructions given in Part 2, and scan your current evening sky for an open cluster. You may do this either by using the **viewing direction** buttons N, S, E, W, and Z on the **button bar**, or by simply "grabbing" the sky with the **hand tool** and dragging while holding the left mouse button down. You may need to zoom in and out a bit to see the dimmer objects. You may also advance time forward to later in the evening if you wish. Read the descriptions given for each object until you find one identifying it as an open cluster. You can confirm that the object is an open cluster by reading what the **object type** is when hovering your mouse over it. Right click (control click for Mac) on this object (being careful to 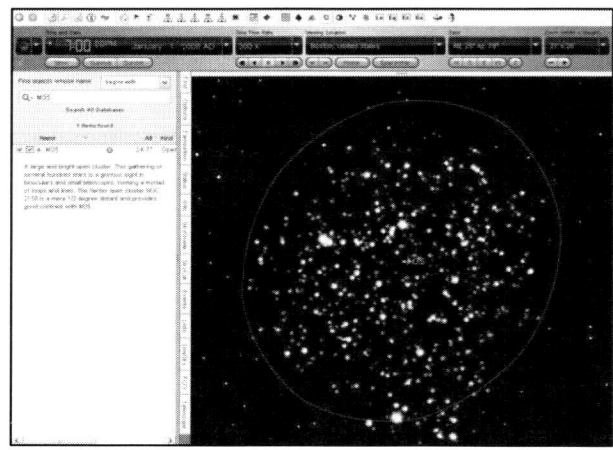 hover the mouse exactly over the object as best you can) and select **centre**. Once centered, zoom in by using the **zoom control** at the right side of the **control panel**. If your chosen object begins to drift from the center of the screen, stop the zooming process and right click (control click for Mac) once again on the object, selecting **centre**.

Once you have zoomed in, right click (control click for Mac) on the object and select **show info** to learn more about it. Expand all layers by clicking the small plus sign to the left (gray arrow for Mac).

Open Cluster #1: Use the information provided by Starry Night to complete the following table:
(Try not to select any of the same examples already given in this activity.)

Date, time:	
Azimuth, altitude:	
Name and catalog number:	
Distance from observer (don't forget units):	
Apparent magnitude:	
Angular size (give units—degrees, arc minutes, etc.):	

Provide a sketch of the object as you see it in Starry Night:

Open Cluster #2: Use the information provided by Starry Night to complete the following table:
(Try not to select any of the same examples already given in this activity.)

Date, time:	
Azimuth, altitude:	
Name and catalog number:	
Distance from observer (don't forget units):	
Apparent magnitude:	
Angular size (give units—degrees, arc minutes, etc.):	

Provide a sketch of the object as you see it in Starry Night:

| |
| |

Open Cluster #3: Use the information provided by Starry Night to complete the following table:
(Try not to select any of the same examples already given in this activity.)

Date, time:	
Azimuth, altitude:	
Name and catalog number:	
Distance from observer (don't forget units):	
Apparent magnitude:	
Angular size (give units—degrees, arc minutes, etc.):	

Provide a sketch of the object as you see it in Starry Night:

| |
| |

PART 5: LOTS AND LOTS OF OPEN CLUSTERS

How many open clusters are out there? Astronomers have only been able to view a small portion of our galaxy. Most of our galaxy is hidden behind dust in the plane of our Milky Way. In fact, we are unable to even see to the center of our own galaxy without the aid of infrared and radio telescopes (longer wavelength radiation passes more easily through interstellar dust). Therefore, of course, the Starry Night databases include only open clusters that have been discovered, recorded, and cataloged. In addition, the Starry Night program only shows objects that you would reasonably be able to see with a small telescope. If you zoom in, however, you simulate viewing conditions through a larger telescope and the program will adjust, showing and labeling additional objects. To get a better sense of spatial distribution, although we can't do anything about the open clusters that are hidden from view and have never been discovered, we can adjust settings to show us all the objects that have been discovered and included in our database.

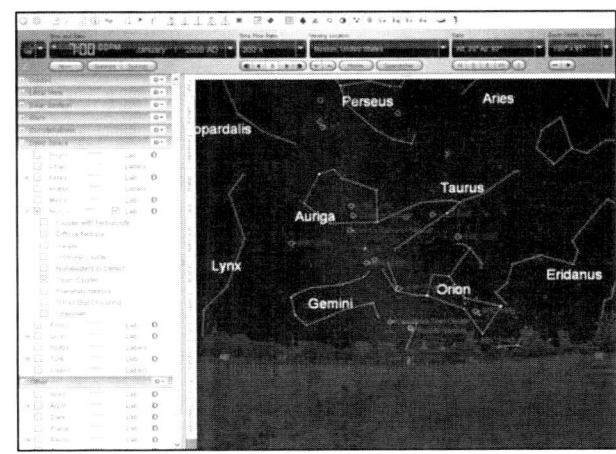

Click the OPTIONS tab on the left **side pane**. Click the plus sign (gray arrow for Mac) to expand the **deep space** and **other** layers. Turn off the **bright NGC objects** and **Messier objects** databases, then click the plus sign (gray arrow for Mac) to expand the **NGC-IC** database and make sure that only the **open cluster** database is turned on. Click the information icon (**i**) to the right to read about this database.

Select the check box on the left and click the **label** box on the right. Now hover your mouse pointer over **NGC-IC** and click to adjust display options. Slide the **number of objects** bar and the **label** bar all the way to the right to maximize the number of objects shown and labeled.

Once you have displayed as many open clusters as you can, use the hand tool to move across the sky to get a sense of the spatial distribution of these objects. Do you notice anything? Do they seem to be concentrated in some areas more than others? Turn on the various coordinate systems by clicking the **coordinate system** buttons* along the **button bar** and see if you can determine what the general alignment is. Can you pose a possible explanation for this? Why don't open clusters appear to be evenly distributed throughout our sky?

*The **lo** button displays a *local coordinate system* based on *azimuth* and *altitude*. The **eq** button displays an *equatorial coordinate system* based on *right ascension* and *declination* (it's similar to Earth's longitude and latitude). The **ec** button displays an *ecliptic coordinate system* keyed to the *ecliptic plane* (the Earth's orbit around the Sun). The **ga** button displays a *galactic coordinate system* keyed to the plane of our galaxy.

Activity 23—GLOBULAR CLUSTERS

*These activities are designed to work with the Starry Night software that comes with your text, from any home location you choose, and with the current date and time, unless indicated otherwise. You may always revert to factory default settings by clicking **FILE/ preferences**, then selecting **factory defaults** as needed. You may also undo a command or series of commands on the PC by clicking the **back** button at the top left of the **button bar**. You should refer to the key given at the beginning of this booklet for clarification of "on screen" buttons, controls, and functions. PC **button bar** items can all be accessed through the **menu**. "Right click" on the PC is equivalent to "control click" on the Mac. All activities assume that OpenGL graphics capabilities are enabled on your computer.*

PART 1: GLOBULAR CLUSTERS

This activity will focus on globular clusters. Globular clusters are densely packed groups of stars that look like cotton balls through a telescope. There may be between 50,000 and 2 million stars in each cluster. The stars tend to be very old and the clusters distributed spherically around the center of our galaxy. Astronomers believe they are remnants from an earlier stage of our galaxy's evolution, before our galaxy flattened out to its current shape.

PART 2: FINDING GLOBULAR CLUSTERS IN OUR NIGHT SKY

You should begin this activity at sunset. An easy way to do this is to click the drop-down menu to the right of the **date & time** field on the **control panel**, and select **sunset**. Look toward the west by clicking the **W viewing direction** button located on the **button bar** across the top of your screen, or by simply keying in the letter W (Mac users should refer to the button bar commands given at the beginning of this booklet). The screen will pan toward the west. Select a playing speed of **300×** normal time by clicking the drop-down menu at the right of the **time speed** field. Click the STOP **time mode** button when the Sun has set, the stars have come out, and dusk is almost over. Then click on the **constellations** button to show the constellations.

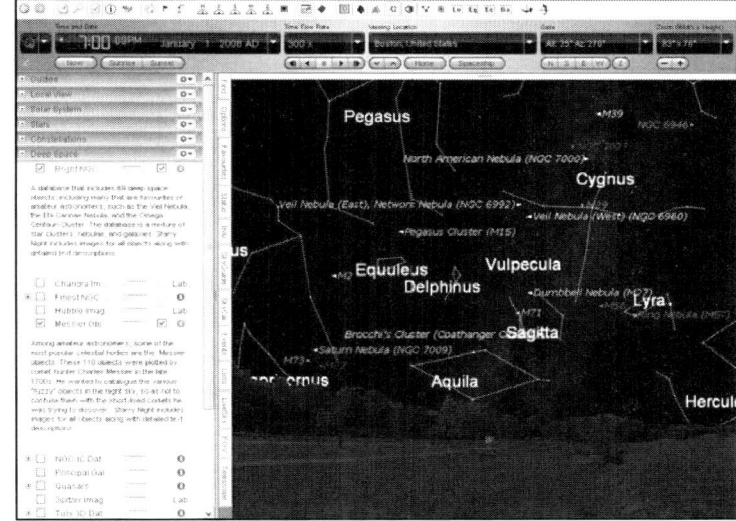

Click the VIEW OPTIONS tab on the left **side pane** and expand the **deep space** layer by clicking on the small plus sign to the left (gray arrow for Mac). Activate the labels for both the **bright NGC objects** and **Messier objects** by clicking the **labels** box to the right. Then click the information icon **(i)** to the far right to display descriptive information on these Starry Night databases. You should now see numerous deep space objects identified on your screen. These are some of the most famous telescope viewing objects in the sky.

[OPTIONAL] For a more complete listing of globular clusters, you may turn on and label all known globular clusters by clicking the box to the left of the **globular cluster** listing in the **stars** layer of the VIEW OPTIONS **side pane**. Don't click the **label** box to the right. A small, green circle will now identify all globular clusters. For this activity, scan the sky only for those that are also labeled, since only these will display an image in Starry Night. Take careful note of the fact that this is an overlapping database. This means that some of the objects in this database are already displayed in Starry Night by other databases. This can lead to confusion since the coordinates may not be exactly the same, thus appearing to indicate that the same object is in two slightly different places at the same time. Although not exact, the additional database markers can help us locate objects. As you zoom in, you should ignore the green circle and look in the general vicinity for the intended object. Once you locate it, right click (control click for Mac) and center it on your screen and continue zooming in. Be careful not to confuse globular clusters with other Messier and NGC objects. The on-screen info will confirm the **object type** when you hover your mouse over it.

PART 3: EXAMPLES OF GLOBULAR CLUSTERS

Let's look at some well-known examples. Astronomers always have a few favorites. Which of these is your favorite?

THE HERCULES CLUSTER (M13):

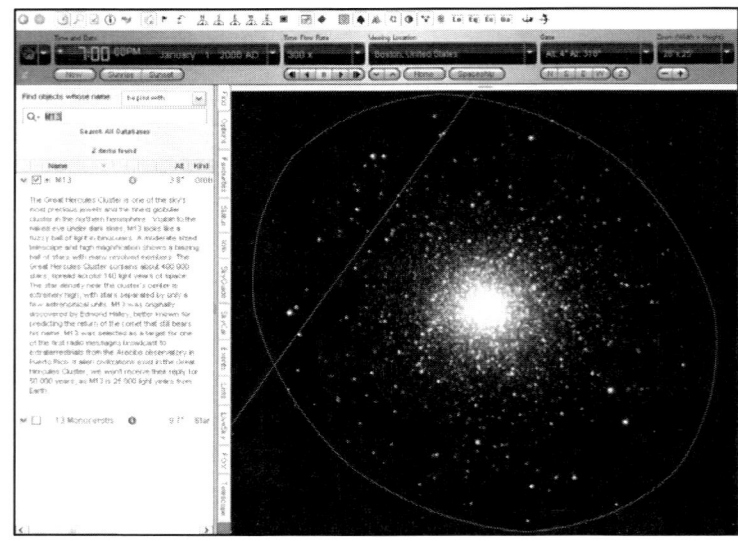

Click the FIND tab on the left **side pane**. In the search field at the top, type in M13. More than one listing may appear. Click the information icon **(i)** to the right for all entries, then grab the right-side double line, and click and drag to the right to expand the text box as far as it will go. You will note that the listings are from different databases. This simply means that Starry Night has this object appearing in multiple (different) databases (or lists). All refer to the same object; however, the information given may be slightly different and the coordinates may vary slightly. Double-click on the listing associated with the **Messier object** database. Zoom in on the object using the **zoom controls** at the right of the **control panel**. Right click (control click for Mac) on the object and select **show info**, click the plus sign (gray arrow for Mac) to expand all layers, then complete the following table of information.

Date, time:	
Azimuth, altitude:	
Distance from observer (give units):	
Apparent magnitude:	
Angular size (give units—degrees, arc minutes, etc.):	

What do you find most interesting? Indicate some additional info about this object:

THE OMEGA CENTAURI CLUSTER (NGC 5139):

Click the FIND tab on the left **side pane**. In the search field at the top, type in NGC 5139. More than one listing may appear. Click the information icon **(i)** to the right for all entries, then grab the right-side double line, and click and drag to the right to expand the text box as far as it will go. You will note that the listings are from different databases. Starry Night has this object appearing in multiple (different) databases (or lists). All refer to the same object; however, the information given may be slightly different and the coordinates may vary slightly as well. Double-click on the listing associated with the **bright NGC objects** database. Adjust the **zoom controls** as needed. Right click (control click for Mac) on the object and select **show info**, click the plus sign (gray arrow for Mac) to expand all layers, then complete the following table of information.

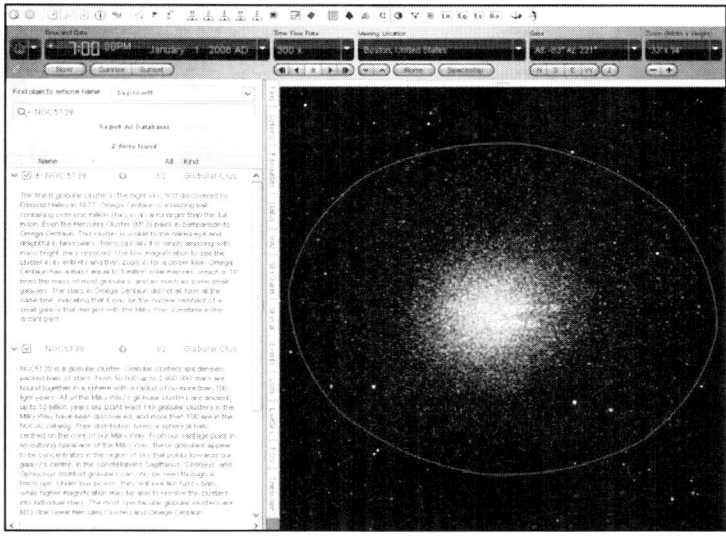

Date, time:	
Azimuth, altitude:	
Distance from observer (give units):	
Apparent magnitude:	
Angular size (give units—degrees, arc minutes, etc.):	

What do you find most interesting? Indicate some additional info about this object:

THE SAGITTARIUS CLUSTER (M22):

Click the FIND tab on the left **side pane**. In the search field at the top, type in M22. More than one listing may appear. Click the information icon **(i)** to the right for all entries, then grab the right-side double line, and click and drag to the right to expand the text box as far as it will go. You will note that the listings are from different databases. Starry Night has this object appearing in multiple (different) databases (or lists). All refer to the same object; however, the information given may be slightly different and the coordinates may vary slightly. Double-click on the listing associated with the **Messier objects** database. Adjust the **zoom controls** as needed. Right click (control click for Mac) on the object and select **show info**, click the plus sign (gray arrow for Mac) to expand all layers, and then complete the following table of information.

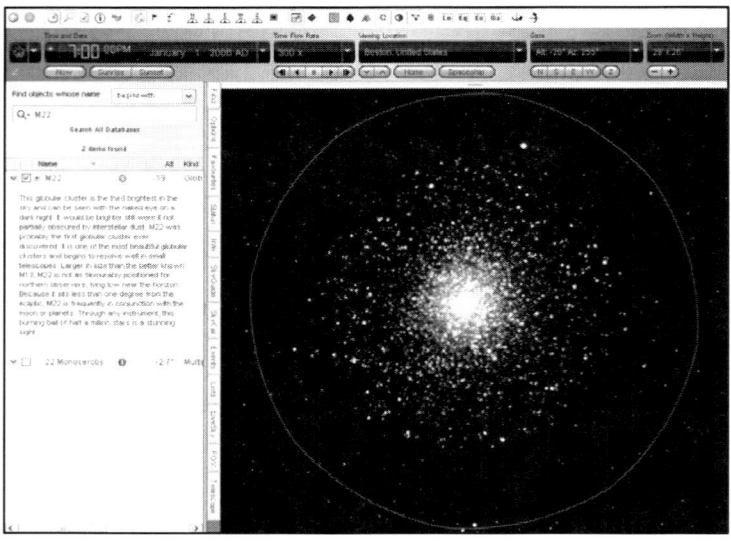

Date, time:	
Azimuth, altitude:	
Distance from observer (give units):	
Apparent magnitude:	
Angular size (give units—degrees, arc minutes, etc.):	

What do you find most interesting? Indicate some additional info about this object:

PART 4: FINDING YOUR OWN EXAMPLES

In this section, we are interested in finding our own examples of globular clusters. Zoom all the way out, return to just after sunset by following the instructions given in Part 2, and scan your current evening sky for a globular cluster. You may do this either by using the **viewing direction** buttons N, S, E, W, and Z on the **button bar**, or by simply "grabbing" the sky with the **hand tool** and dragging while holding the left mouse button down. You may need to zoom in and out a bit to see the dimmer objects. You may also advance time forward to later in the evening if you wish. Read the descriptions given for each object until you find one identifying it as a globular cluster. You can confirm that the object is a globular cluster by reading what the **object type** is when hovering your mouse over it. Right click (control click for Mac) on this object (being careful to hover the mouse exactly over the object as best you can) and select **centre**. Once centered, zoom in by using the **zoom control** at the right side of the **control panel**. If your chosen object begins to drift from the center of the screen, stop the zooming process and right click (control click for Mac) once again on the object, selecting **centre**.

Once you have zoomed in, right click (control click for Mac) on the object and select **show info** to learn more about it. Expand all layers by clicking the small plus sign to the left (gray arrow for Mac).

Globular Cluster #1: Use the information provided by Starry Night to complete the following table:
(Try not to select any of the same examples already given in this activity.)

Date, time:	
Azimuth, altitude:	
Name and catalog number:	
Distance from observer (don't forget units):	
Apparent magnitude:	
Angular size (give units—degrees, arc minutes, etc.):	

Provide a sketch of the object as you see it in Starry Night:

Globular Cluster #2: Use the information provided by Starry Night to complete the following table:
(Try not to select any of the same examples already given in this activity.)

Date, time:	
Azimuth, altitude:	
Name and catalog number:	
Distance from observer (don't forget units):	
Apparent magnitude:	
Angular size (give units—degrees, arc minutes, etc.):	

Provide a sketch of the object as you see it in Starry Night:

<u>Globular Cluster #3:</u> Use the information provided by Starry Night to complete the following table:
(Try not to select any of the same examples already given in this activity.)

Date, time:	
Azimuth, altitude:	
Name and catalog number:	
Distance from observer (don't forget units):	
Apparent magnitude:	
Angular size (give units—degrees, arc minutes, etc.):	

Provide a sketch of the object as you see it in Starry Night:

PART 5: LOTS AND LOTS OF GLOBULAR CLUSTERS

How many globular clusters are out there? Astronomers have only been able to view a small portion of our galaxy. Most of our galaxy is hidden behind dust in the plane of our Milky Way. In fact, we are unable to even see to the center of our own galaxy without the aid of infrared and radio telescopes (longer wavelength radiation passes more easily through interstellar dust). Therefore, the Starry Night databases include only globular clusters that have been discovered, recorded, and cataloged. In addition, the Starry Night program only shows objects that you would reasonably be able to see with a small telescope. If you zoom in, however, you simulate viewing conditions through a larger telescope and the program will adjust, showing and labeling additional objects. To get a better sense of spatial distribution, although we can't do anything about the globular clusters that are hidden from view and have never been discovered, we can adjust settings to show us all the objects that have been discovered and included in our database.

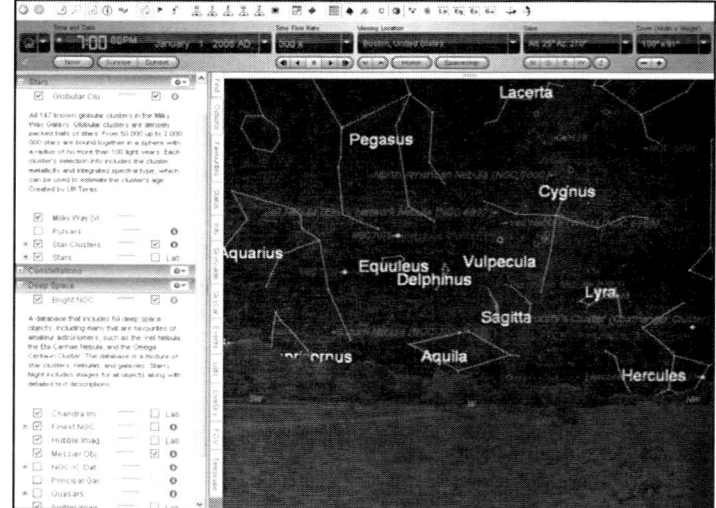

Click the OPTIONS tab on the left **side pane**. Click the plus sign (gray arrow for Mac) to expand the **stars** layer. Find the **globular cluster** database and click the information icon **(i)** to the right to read about this database. Select the check box on the left and click the **label** box on the right. Now hover your mouse pointer over the words **globular cluster** and click to adjust display options. Slide the **number of objects** bar and the **label** bar all the way to the right to maximize the number of objects shown and labeled.

Once you have displayed as many globular clusters as you can, use the hand tool to move across the sky to get a sense of the spatial distribution of these objects. Do you notice anything? Do they seem to be concentrated in some areas more than others? Turn on the various coordinate systems by clicking the **coordinate system** buttons* along the **button bar** and see if you can determine what the general alignment is. Can you pose a possible explanation for this? Try and explain why globular clusters do not appear to be evenly distributed throughout our sky as one might imagine due to their known spherical distribution around the centers of our galaxies?

*The **lo** button displays a *local coordinate system* based on *azimuth* and *altitude*. The **eq** button displays an *equatorial coordinate system* based on *right ascension* and *declination* (it's similar to Earth's longitude and latitude). The **ec** button displays an *ecliptic coordinate system* keyed to the *ecliptic plane* (the Earth's orbit around the Sun). The **ga** button displays a *galactic coordinate system* keyed to the plane of our galaxy.

PART 6: DISTRIBUTION, AGE, AND SPECTRAL CHARACTERISTICS OF CLUSTERS:
[OPTIONAL—*for more advanced classes*]

Be sure that the **globular cluster** database is turned on and labeled (as described in the previous section), while the **Messier object** and **bright NGC objects** databases are turned off (click the OPTIONS tab on the left **side pane**, click the plus sign (gray arrow for Mac) to expand the **deep space** layer, and uncheck the **bright NGC objects** and **Messier objects** databases). Select a random sample of 5-10 globular clusters. Without zooming in, right click (control click for Mac) on the cluster and select **show info**, then complete the following table.

NAME (NGC #)	Dist from Gal. center	Absolute Vis. Magn.	Int. Spectral Type	Dist from Our Sun	Horizontal Branch Type	Metallicity [Fe/He]

Your lecture course instructor will discuss the significance of some of these parameters as they relate to the physical characteristics and age of the globular cluster. The Distance from the Galactic Center, Absolute Visual Magnitude, and Distance from Our Sun, help us determine the special distribution of globular clusters around our galaxy. The Int. Spectral Type, Horizontal Branch Type, and Metallicity are important parameters that indicate the age of the globular cluster. Globular clusters are thought to have formed in the early history of our galaxy's formation, when our galaxy had a spherical distribution. As our galaxy flattened out, globular clusters remained in the original spherical distribution of the early galaxy and slowly oscillated through the central region (rather than revolve around the central region). Stripped of their interstellar dust and gasses, clusters are mostly comprised of old and dying stars. They have much to tell us of the conditions of the early universe and may hold secrets to our own galaxies formation and evolution.

Activity 24—PLANETARY NEBULAE

*These activities are designed to work with the Starry Night software that comes with your text, from any home location you choose, and with the current date and time, unless indicated otherwise. You may always revert to factory default settings by clicking **FILE / preferences**, then selecting **factory defaults** as needed. You may also undo a command or series of commands on the PC by clicking the **back** button at the top left of the **button bar**. You should refer to the key given at the beginning of this booklet for clarification of "on screen" buttons, controls, and functions. PC **button bar** items can all be accessed through the **menu**. "Right click" on the PC is equivalent to "control click" on the Mac. All activities assume that OpenGL graphics capabilities are enabled on your computer.*

PART 1: NEBULAE

The term *nebulae* initially included all "fuzzy" objects seen in telescopes, but now many of them have been identified as galaxies, open clusters, globular clusters, stellar associations, supernova remnants, planetary nebulae, emission nebulae, reflection nebulae, and dark nebulae. This activity will focus on planetary nebulae.

Planetary nebulae are formed when a relatively small star, such as our Sun, runs out of fuel. The core of the star collapses to form a very hot and dense white dwarf star. An outer shell of gas is "blown off" during the collapse, producing the spectacular "ring" or "shell" that we see. Try to identify the central white dwarf. Although they are hot and bright, their small size limits our ability to see them. For example, a star the size of our Sun will collapse down to a white dwarf about the size of the Earth, a hundredth the diameter and only one-millionth its current volume. Although the size of the star is greatly reduced, the mass stays about the same (except for the gasses ejected into the shell), so white dwarfs are incredibly dense objects, over a million times more dense than that of our Sun.

You may see different colors. The different colors of the gasses may represent different elements depending on the source of the image, filters used, and digital image processing techniques employed.

PART 2: FINDING PLANETARY NEBULAE IN OUR NIGHT SKY

You should begin this activity at sunset. An easy way to do this is to click the drop-down menu to the right of the **date & time** field on the **control panel**, and select **sunset**. Look toward the west by clicking the **W viewing direction** button located on the **button bar** across the top of your screen, or by simply keying in the letter W (Mac users should refer to the button bar commands given at the beginning of this booklet). The screen will pan toward the west. Select a playing speed of **300×** normal time by clicking the drop-down menu at the right of the **time speed** field. Click the STOP **time mode** button when the Sun has set, the stars have come out, and dusk is almost over. Then click on the **constellations** button to show the constellations.

Click the OPTIONS tab on the left **side pane** and expand the **deep space** layer by clicking on the small plus sign to the left (gray arrow for Mac). Activate the labels for both the **bright NGC objects** and **Messier objects** by clicking the **labels** box to the right. Then click the information icon **(i)** to the far right to display descriptive information on these Starry Night databases. You should now see numerous deep space objects identified on your screen. These are some of the most famous telescope viewing objects in the sky.

[OPTIONAL] For a more complete listing of planetary nebulae, you may turn on the planetary nebula database by clicking the box to the left of the **planetary nebula** listing in the **other** layer of the OPTIONS **side pane**. Don't click the **label** box to the right. A small, green circle with crosshairs will now identify all planetary nebulae. For this activity, scan the sky only for those that are also labeled, since only these will display an image in Starry Night™ Pro. Take careful note of the fact that this is an overlapping database. This means that some of the objects in this database are already displayed in Starry Night by other databases. This can lead to confusion since the coordinates may not be exactly the same, thus appearing to indicate that the same object is in two slightly different places at the same time. Although not exact, the additional database markers can help us locate objects. As you zoom in, you should ignore the green circle with crosshairs and look in the general vicinity for the intended object. Once you locate it, right click (control click for Mac) and center it on your screen and continue zooming in. Be careful not to confuse planetary nebulae with other Messier and NGC objects. The on-screen info will confirm the **object type** when you hover your mouse over it.

PART 3: EXAMPLES OF PLANETARY NEBULAE

Let's look at some well-known examples. Astronomers always have a few favorites. Which of these is your favorite?

THE RING NEBULA (M57):

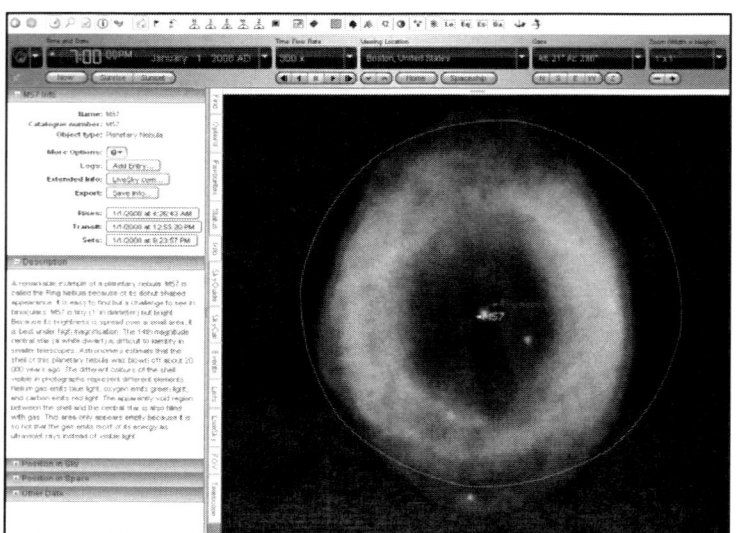

Click the FIND tab on the left **side pane**. In the search field at the top, type in RING NEBULA. Two listings will appear. Grab the right-side double line, click and drag to the right to expand the text box. Click the information icon (**i**) to the right for both entries. You will note that the first listing is from the **NGC-IC** database, the second listing is from the **Messier object** database. This simply means that Starry Night has this object appearing twice, on two different databases (or lists). Both listings refer to the same object; however, the information given may be slightly different and the coordinates may be slightly off as well. Double-click on the second of the two, the listing associated with the **Messier object** database. This will give us more detailed and specific information on the Ring Nebula. Right click (control click for Mac) on the object and select **show info**, click the plus sign (gray arrow for Mac) to expand all layers, then complete the following table of information.

Date, time:	
Azimuth, altitude:	
Distance from observer (give units):	
Apparent magnitude:	
Angular size (give units—degrees, arc minutes, etc.):	

What do you find most interesting? Indicate some additional info about this object:

THE DUMBBELL NEBULA (M27):

Click the FIND tab on the left **side pane**. In the search field at the top, type in DUMBBELL NEBULA. Double-click to select this item and center it. You may need to adjust the Zoom Control a bit. Right click (control click for Mac) on the object and select INFO, click the plus sign (gray arrow for Mac) to expand all layers, then complete the following table of information.

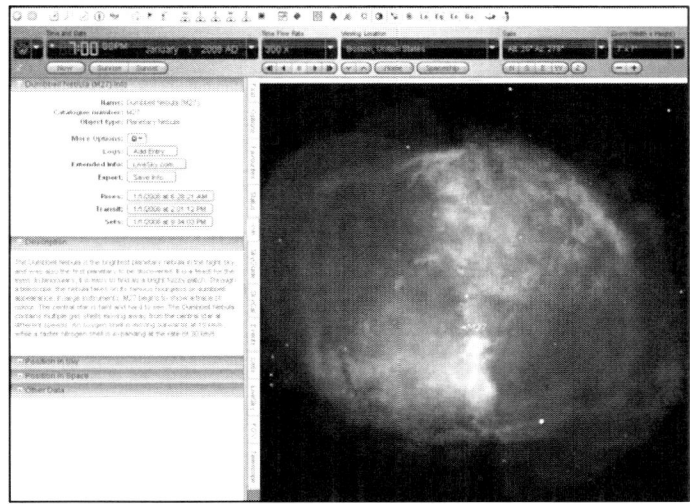

Date, time:	
Azimuth, altitude:	
Distance from observer (give units):	
Apparent magnitude:	
Angular size (give units—degrees, arc minutes, etc.):	

What do you find most interesting? Indicate some additional info about this object:

THE ESKIMO NEBULA or CLOWN NEBULA (NGC 2392):

Click the FIND tab on the left **side pane**. In the search field at the top, type in ESKIMO NEBULA. Two listings will appear. Grab the right-side double line, click and drag to the right to expand the text box. Click the information icon **(i)** to the right for both entries. You will note that the first listing is from the **Herschel 400** database; the second listing is from the **SAA 100** database. Starry Night has this object appearing twice, in two different databases. Both listings refer to the same object; however, the information given may be slightly different and the coordinates may be slightly off as well. Double-click on the second of the two, the listing associated with the **SAA 100** database. This will give us more detailed and specific information on the Eskimo Nebula. Right

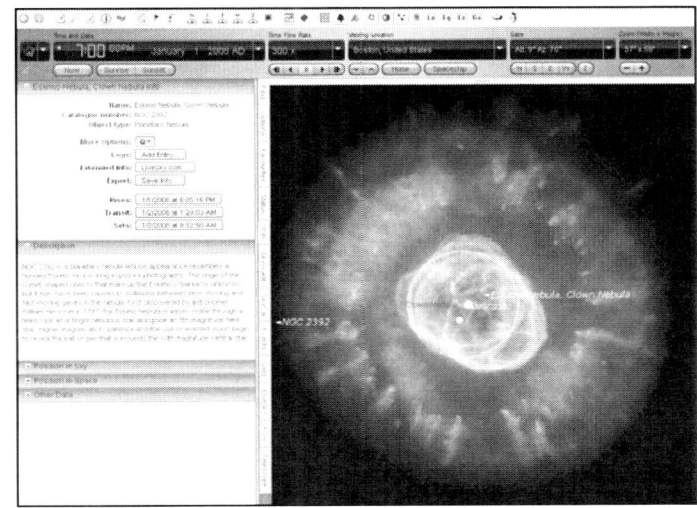

click (control click for Mac) on the object and select **show info**, click the plus sign (gray arrow for Mac) to expand all layers, then complete the following table of information.

Date, time:	
Azimuth, altitude:	
Distance from observer (give units):	
Apparent magnitude:	
Angular size (give units—degrees, arc minutes, etc.):	

What do you find most interesting? Indicate some additional info about this object:

THE GHOST of JUPITER or EYE NEBULA (NGC 3242):

Click the FIND tab on the left **side pane**. In the search field at the top, type in NGC 3242. More than one listing may appear. Click the information icon **(i)** to the right for all entries, then grab the right-side double line, and click and drag to the right to expand the text box as far as it will go. You will note that the listings are from different databases. Starry Night has this object appearing in multiple (different) databases (or lists). All refer to the same object; however, the information given may be slightly different and the coordinates may vary slightly. Double-click on the listing associated with the **bright NGC objects** database. Adjust the **zoom controls** as needed. Right click (control click for Mac) on the object and select **show info**, click the plus sign (gray arrow for Mac) to expand all layers, and then complete the following table of information.

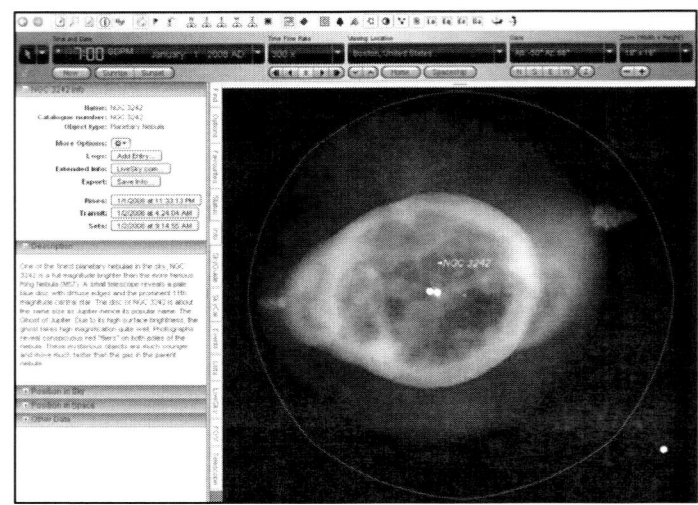

Date, time:	
Azimuth, altitude:	
Distance from observer (give units):	
Apparent magnitude:	
Angular size (give units -- degrees, arc minutes, etc.):	

What do you find most interesting? Indicate some additional info about this object:

PART 4: FINDING YOUR OWN EXAMPLES

In this section, we are interested in finding our own examples of planetary nebulae. Zoom all the way out, return to just after sunset by following the instructions given in Part 2, and scan your current evening sky for a planetary nebula. You may do this either by using the direction buttons N, S, E, W, and Z on the **button bar**, or by simply "grabbing" the sky with the **hand tool** and dragging while holding the left mouse button down. You may need to zoom in and out a bit to see the dimmer objects. You may also advance time forward to later in the evening if you wish. Read the descriptions given for each object until you find one identifying it as a planetary nebula. You can confirm that the object is a planetary nebula by reading what the **object type** is when hovering your mouse over it. Right click (control click for Mac) on this object

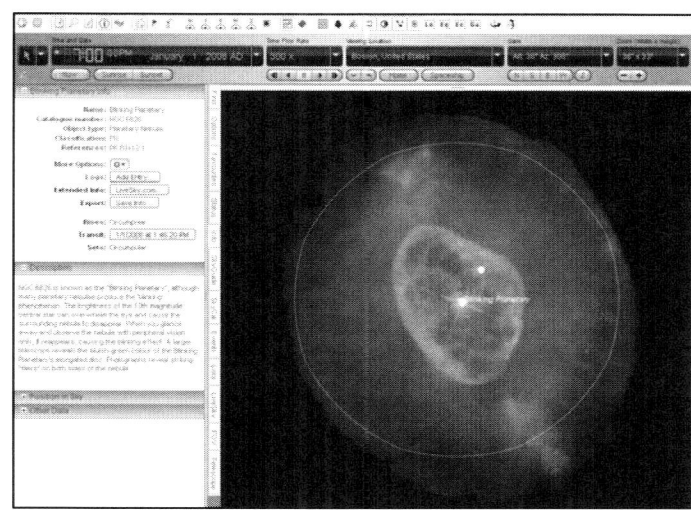

(being careful to hover the mouse exactly over the object as best you can) and select **centre**. Once centered, zoom in by using the **zoom control** at the right side of the **control panel**. If your chosen object begins to drift from the center of the screen, stop the zooming process and right click (control click for Mac) once again on the object, selecting **centre**.

Once you have zoomed in, right click (control click for Mac) on the object and select **show info** to learn more about it. Click the plus sign (gray arrow for Mac) to expand all layers.

Planetary Nebula #1: Use the information provided by Starry Night to complete the following table:
(Try not to select any of the same examples already given in this activity.)

Date, time:	
Azimuth, altitude:	
Name and catalog number:	
Distance from observer (give units):	
Apparent magnitude:	
Angular size (give units -- degrees, arc minutes, etc.):	

Provide a sketch of the object as you see it in Starry Night:

<u>Planetary Nebula #2:</u> Use the information provided by Starry Night to complete the following table:
(Try not to select any of the same examples already given in this activity.)

Date, time:	
Azimuth, altitude:	
Name and catalog number:	
Distance from observer (give units):	
Apparent magnitude:	
Angular size (give units -- degrees, arc minutes, etc.):	

Provide a sketch of the object as you see it in Starry Night:

<u>Planetary Nebula #3:</u> Use the information provided by Starry Night to complete the following table:
(Try not to select any of the same examples already given in this activity.)

Date, time:	
Azimuth, altitude:	
Name and catalog number:	
Distance from observer (give units):	
Apparent magnitude:	
Angular size (give units -- degrees, arc minutes, etc.):	

Provide a sketch of the object as you see it in Starry Night:

PART 5: LOTS AND LOTS OF PLANETARY NEBULAE

How many planetary nebulae are out there? Astronomers have only been able to view a small portion of our galaxy. Most of our galaxy is hidden behind dust in the plane of our Milky Way. In fact, we are unable to even see to the center of our own galaxy without the aid of infrared and radio telescopes (longer wavelength radiation passes more easily through interstellar dust). Therefore, the Starry Night databases include only planetary nebulae that have been discovered, recorded, and cataloged. In addition, the Starry Night program only shows objects that you would reasonably be able to see with a small telescope. If you zoom in, however, you simulate viewing conditions through a larger telescope and the program will adjust, showing and labeling additional objects. To get a better sense of spatial distribution, although we can't do anything about the planetary nebulae which are hidden from view and have never been discovered, we can adjust settings to show us all the objects that have been discovered and included in our database.

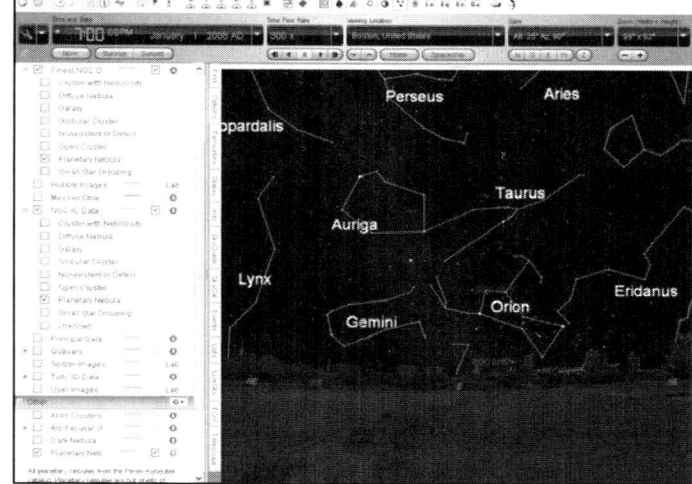

Click the OPTIONS tab on the left **side pane**. Click the plus sign (gray arrow for Mac) to expand the **other** layer. Find the **planetary nebula** database and click the information icon **(i)** to the right to read about this database. Select the check box on the left and click the **label** box on the right. Now hover your mouse pointer over the words **planetary nebula** and click to adjust display options. Slide the **number of objects** bar and the **label** bar all the way to the right to maximize the number of objects shown and labeled.

You can turn on other databases that include planetary nebulae as well. For example, the **Finest NGC** and **NGC-IC** databases each give you the option to select and display only planetary nebulae. You should note that many of these databases will overlap and list the same object multiple times. There may be slight variations in the coordinates, giving multiple views of the same object. Don't let this confuse you. Simply turn off all other databases and your object will be uniquely identified.

Once you have displayed as many planetary nebulae as you can, use the hand tool to move across the sky to get a sense of the spatial distribution of these objects. Do you notice anything? Do they seem to be concentrated in some areas more than others? Turn on the various coordinate systems by clicking the **coordinate system** buttons* along the **button bar** and see if you can determine what the general alignment is. Can you pose a possible explanation for this?

*The **lo** button displays a *local coordinate system* based on *azimuth* and *altitude*. The **eq** button displays an *equatorial coordinate system* based on *right ascension* and *declination* (it's similar to Earth's longitude and latitude). The **ec** button displays an *ecliptic coordinate system* keyed to the *ecliptic plane* (the Earth's orbit around the Sun). The **ga** button displays a *galactic coordinate system* keyed to the plane of our galaxy.

Activity 25—EMISSION NEBULAE

*These activities are designed to work with the Starry Night software that comes with your text, from any home location you choose, and with the current date and time, unless indicated otherwise. You may always revert to factory default settings by clicking **FILE/ preferences**, then selecting **factory defaults** as needed. You may also undo a command or series of commands on the PC by clicking the **back** button at the top left of the **button bar**. You should refer to the key given at the beginning of this booklet for clarification of "on screen" buttons, controls, and functions. PC **button bar** items can all be accessed through the **menu**. "Right click" on the PC is equivalent to "control click" on the Mac. All activities assume that OpenGL graphics capabilities are enabled on your computer.*

PART 1: NEBULAE

The term *nebulae* initially included all "fuzzy" objects seen in telescopes, but now many of them have been identified as galaxies, open clusters, globular clusters, stellar associations, supernova remnants, planetary nebulae, emission nebulae, reflection nebulae, and dark nebulae. This activity will focus on emission nebulae.

The different colors of the gasses may represent different elements, depending on the source of the image, filters used, and digital image processing techniques employed. Often, reflection nebulae accompany emission nebulae as with the Trifid Nebula (M20), where a blue reflection nebula surrounds a red emission nebula. In this case, the red color is due to hydrogen emission, while the blue color is due to reflection caused by dust particles. Dark nebulae are produced when the dust is so thick that it blocks light from background stars. It's paradoxical to note that these dark areas are often intense star-forming regions.

PART 2: FINDING NEBULAE IN OUR NIGHT SKY

You should begin this activity at sunset. An easy way to do this is to click the drop-down menu to the right of the **date & time** field on the **control panel**, and select **sunset**. Look toward the west by clicking the **W viewing direction** button located on the **button bar** across the top of your screen, or by simply keying in the letter W (Mac users should refer to the button bar commands given at the beginning of this booklet). The screen will pan toward the west. Select a playing speed of **300×** normal time by clicking the drop-down menu at the right of the **time speed** field. Click the STOP **time mode** button when the Sun has set, the stars have come out, and dusk is almost over. Then click on the **constellations** button to show the constellations.

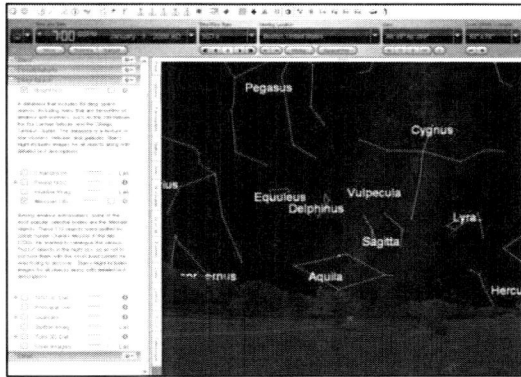

Click the OPTIONS tab on the left **side pane** and expand the **deep space** layer by clicking on the small plus sign to the left (gray arrow for Mac). Activate the labels for both the **bright NGC objects** and **Messier objects** by clicking the **labels** box to the right. Then click the information icon **(i)** to the far right to display descriptive information on these Starry Night databases.

You should now see numerous deep-space objects identified on your screen.

PART 3: EXAMPLES OF EMISSION NEBULAE

Let's look at some well-known examples. Astronomers always have a few favorites. Which of these is your favorite?

THE ORION & HORSEHEAD NEBULAE (M42 & NGC 2024):

Click the FIND tab on the left **side pane**. In the search field at the top, type in M42. More than one listing may appear. Click the information icon **(i)** to the right for all entries, then grab the right-side double line, and click and drag to the right to expand the text box as far as it will go. You will note that the listings are from different databases. This simply means that Starry Night has this object appearing in multiple (different) databases (or lists). All refer to the same object; however, the information given may be slightly different, and the coordinates may vary slightly as well. Double click on the listing associated with the **Messier object** database.

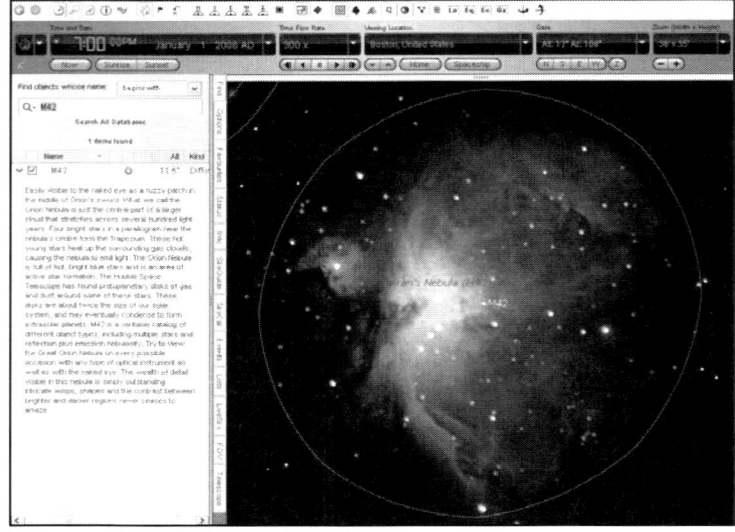

Zoom in on the object using the **zoom controls** at the right of the **control panel**. Right click (control click for Mac) on the object and select **show info**, click the plus sign (gray arrow for Mac) to expand all layers, then complete the following table of information. To see the Horsehead Nebula, you will need to zoom out a bit to find it, then recenter, and zoom back in.

Name and catalog number:	**M42 (Orion Nebula)**	**NGC 2024 (Horsehead)**
Date, time:		
Azimuth, altitude:		
Distance from observer (give units):		
Apparent magnitude:		
Angular size (give units—degrees, arc minutes, etc.):		

What do you find most interesting? Indicate some additional info about these objects:

THE LAGOON & TRIFID NEBULAE (M8 & M20):

Click the FIND tab on the left **side pane**. In the search field at the top, type in M8. More than one listing may appear. Click the information icon **(i)** to the right for all entries, then grab the right-side double line and click and drag to the right to expand the text box as far as it will go. You will note that the listings are from different databases. Starry Night has this object appearing in multiple (different) databases (or lists). All refer to the same object; however, the information given may be slightly different, and the coordinates may vary slightly as well. Double-click on the listing associated with the **Messier objects** database. Adjust the **zoom controls** as needed. Right click (control click for Mac) on the object and select **show info**, click the plus sign (gray arrow for Mac) to expand all layers, then complete the following table of information. To see the Trifid nebula, you will need to zoom out a bit to find it, then recenter, and zoom back in.

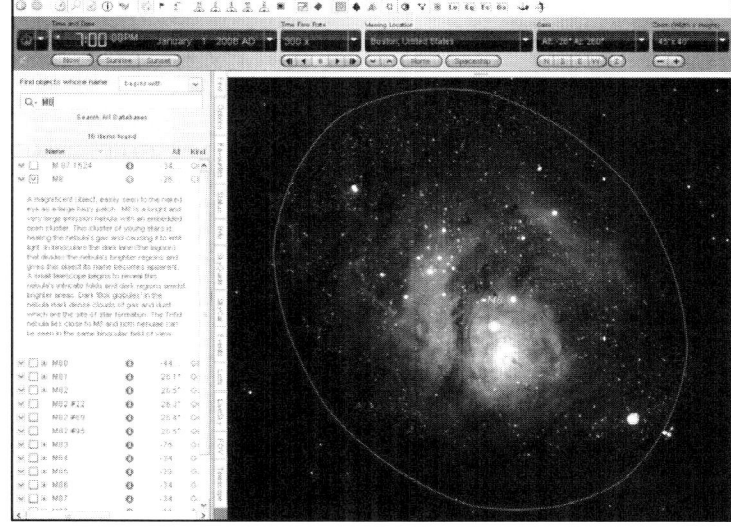

Name and catalog number:	**M8 (Lagoon Nebula)**	**M20 (Trifid Nebula)**
Date, time:		
Azimuth, altitude:		
Distance from observer (give units):		
Apparent magnitude:		
Angular size (give units—degrees, arc minutes, etc):		

What do you find most interesting? Indicate some additional info about these objects:

THE KEYHOLE NEBULA (NGC 3372):

Click the FIND tab on the left **side pane**. In the search field at the top, type in NGC 3372. More than one listing may appear. Click the information icon **(i)** to the right for all entries, then grab the right-side double line and click and drag to the right to expand the text box as far as it will go. You will note that the listings are from different databases. Starry Night has this object appearing in multiple (different) databases (or lists). All refer to the same object; however, the information given may be slightly different, and the coordinates may vary slightly as well. Double-click on the listing associated with the **bright NGC objects** database. Adjust the **zoom controls** as needed. Right click (control click for Mac) on the object and select **show info**, click the plus sign (gray arrow for Mac) to expand all layers, then complete the following table of information.

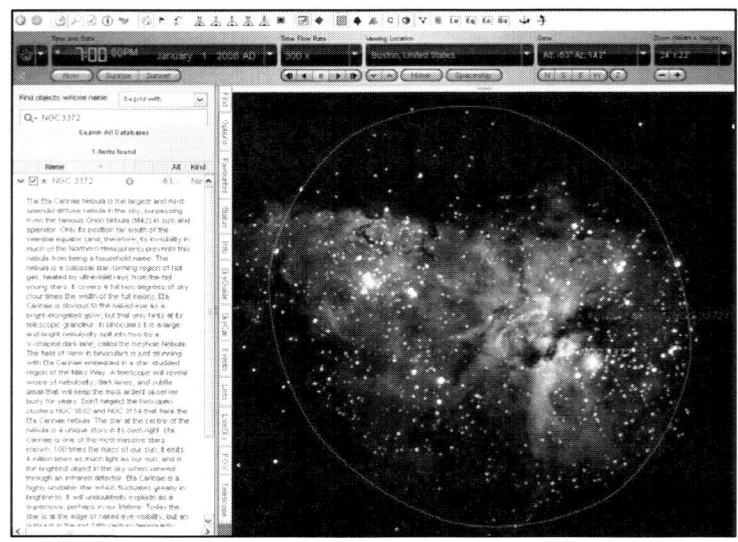

Date, time:	
Azimuth, altitude:	
Distance from observer (give units):	
Apparent magnitude:	
Angular size (give units—degrees, arc minutes, etc.):	

What do you find most interesting? Indicate some additional info about this object:

PART 4: FINDING YOUR OWN EXAMPLES

In this section, we are interested in finding our own examples of emission nebulae: diffuse clouds of gas and dust that emit light due to high-energy interactions with nearby stars. Zoom all the way out, return to just after sunset following the instructions given in Part 2, and scan your current evening sky for a nebula. You may do this either by using the **viewing direction** buttons N, S, E, W, and Z on the **button bar**, or by simply "grabbing" the sky with the **hand tool** and dragging while holding the left mouse button down. You may need to zoom in and out a bit to see the dimmer objects. You may also advance time forward to later in the evening if you wish. Read the descriptions given for each object until you find one identifying it as a nebula. You can confirm that the object is an emission nebula by reading what the **object type** is when hovering your mouse over it.

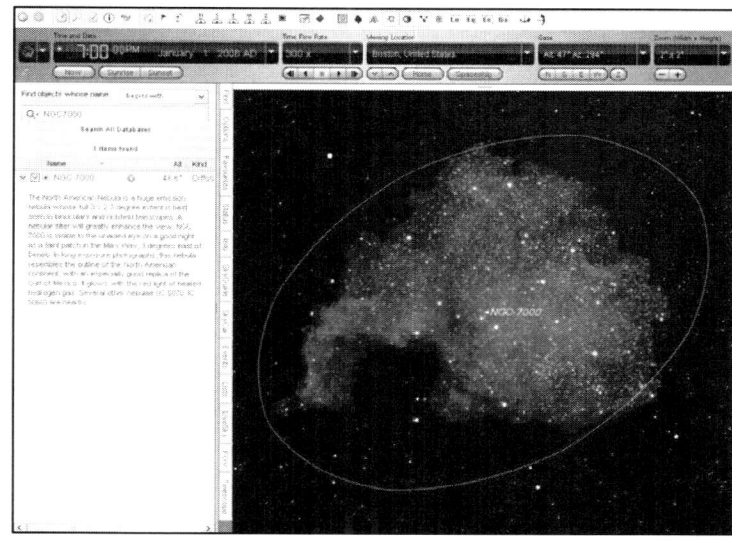

Right click (control click for Mac) on this object (careful to hover the mouse exactly over the object as best you can) and select **centre**. Once centered, zoom in by using the **zoom control** at the right side of the **control panel**. If your chosen object begins to drift from the center of the screen, stop the zooming process and right click (control click for Mac) once again on the object, selecting **centre**.

Once you have zoomed in, right click (control click for Mac) on the object and select **show info** to learn more about it. Click the plus sign (gray arrow for Mac) to expand all layers.

Nebula #1: Use the information provided by Starry Night to complete the following table:
(Try not to select any of the same examples already given in this activity.)

Date, time:	
Azimuth, altitude:	
Name and catalog number:	
Distance from observer (give units):	
Apparent magnitude:	
Angular size (give units—degrees, arc minutes, etc.):	

Provide a sketch of the object as you see it in Starry Night:

Nebula #2: Use the information provided by Starry Night to complete the following table:
(Try not to select any of the same examples already given in this activity.)

Date, time:	
Azimuth, altitude:	
Name and catalog number:	
Distance from observer (give units):	
Apparent magnitude:	
Angular size (give units—degrees, arc minutes, etc.):	

Provide a sketch of the object as you see it in Starry Night:

Nebula #3: Use the information provided by Starry Night to complete the following table:
(Try not to select any of the same examples already given in this activity.)

Date, time:	
Azimuth, altitude:	
Name and catalog number:	
Distance from observer (give units):	
Apparent magnitude:	
Angular size (give units—degrees, arc minutes, etc.):	

Provide a sketch of the object as you see it in Starry Night:

Activity 26—SUPERNOVA REMNANTS

*These activities are designed to work with the Starry Night software that comes with your text, from any home location you choose, and with the current date and time, unless indicated otherwise. You may always revert to factory default settings by clicking **FILE/preferences**, then selecting **factory defaults** as needed. You may also undo a command or series of commands on the PC by clicking the **back** button at the top left of the **button bar**. You should refer to the key given at the beginning of this booklet for clarification of "on screen" buttons, controls, and functions. PC **button bar** items can all be accessed through the **menu**. "Right click" on the PC is equivalent to "control click" on the Mac. All activities assume that OpenGL graphics capabilities are enabled on your computer.*

PART 1: SUPERNOVA

A supernova will occur when a massive star runs out of fuel and becomes unstable, resulting in a powerful explosion that can cause the star to temporarily outshine its host galaxy. When these occur in our own galaxy, they may shine as bright as the Moon and be seen with the naked eye in daytime for weeks and at night for years. Eventually, they dissipate into the interstellar medium, sometimes leaving behind a supernova remnant. Although some supernova remnants are visible optically, many can be observed in other regions of the electromagnetic spectrum. Certain types of supernova may leave behind black holes (detected by X-rays) or spinning neutron stars called pulsars. Thus X-ray sources and pulsars may also be indicators of supernovae.

PART 2: FINDING SUPERNOVA REMNANTS IN OUR NIGHT SKY

You should begin this activity at sunset. An easy way to do this is to click the drop-down menu to the right of the **date & time** field on the **control panel**, and select **sunset**. Look toward the west by clicking the **W viewing direction** button located on the **button bar** across the top of your screen, or by simply keying in the letter W (Mac users should refer to the button bar commands given at the beginning of this booklet). The screen will pan toward the west. Select a playing speed of **300×** normal time by clicking the drop-down menu at the right of the **time speed** field. Click the STOP **time mode** button when the Sun has set, the stars have come out, and dusk is almost over. Then click on the **constellations** button to show the constellations.

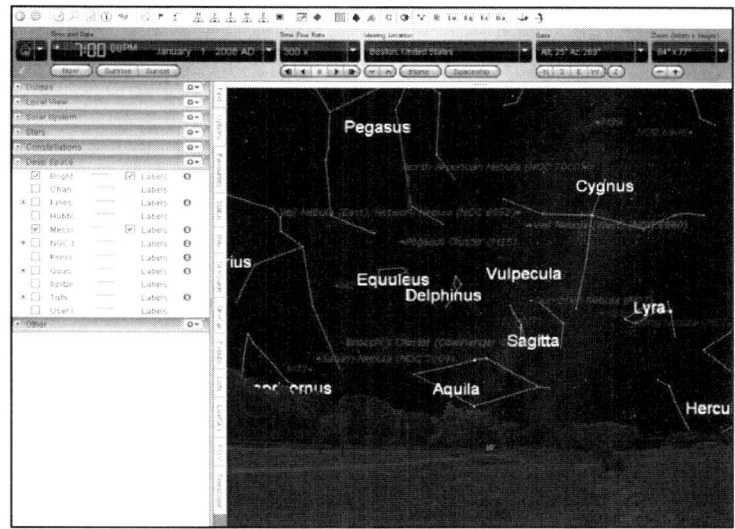

Click the VIEW OPTIONS tab on the left **side pane** and expand the **deep space** layer by clicking on the small plus sign to the left (gray arrow for Mac). Activate the labels for both the **bright NGC objects** and **Messier objects** by clicking the **labels** box to the right. Then click the information icon **(i)** to the far right to display descriptive information on these Starry Night databases.

You should now see numerous deep-space objects identified on your screen.

In this activity, we are interested in finding supernova remnants. Here are a few of the most famous supernova remnants visible optically through telescopes.

THE CRAB NEBULA (M1):

Click the FIND tab on the left **side pane**. In the search field at the top, type in M1. More than one listing may appear. Click the Information icon **(i)** to the right for all entries, then grab the right-side double line and click and drag to the right to expand the text box as far as it will go. You will note that the listings are from different databases. This simply means that Starry Night has this object appearing in multiple (different) databases (or lists). All refer to the same object; however, the information given may be slightly different, and the coordinates may vary slightly as well. Double-click on the listing associated with the **Messier object** database. Zoom in on the object using the **zoom controls** at the right of the **control panel**. Right click (control click for Mac) on the object and select **show info**, click the plus sign (gray arrow for Mac) to expand all layers, and then complete the following table of information.

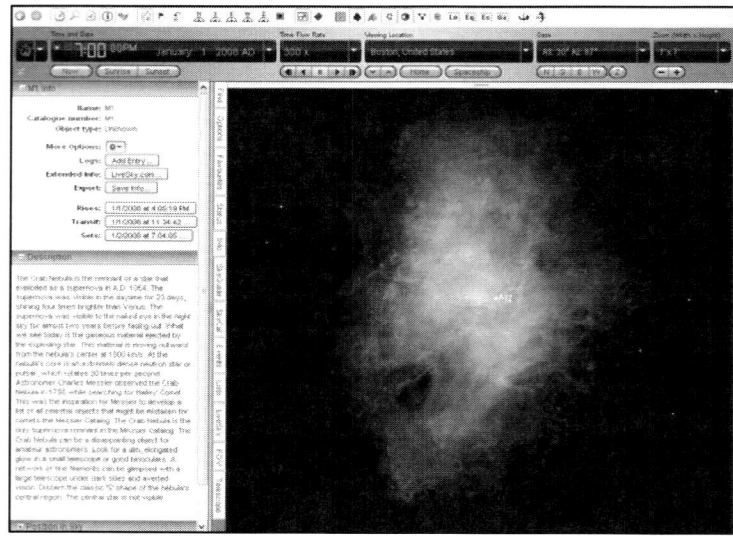

Date, time:	
Azimuth, altitude:	
Distance from observer (give units):	
Apparent magnitude:	
Angular size (give units—degrees, arc minutes, etc.):	

What do you find most interesting? Indicate some additional info about this object:

THE VEIL NEBULA (NGC 6992 & NGC 6960):

Click the FIND tab on the left **side pane**. In the search field at the top, type in NGC 6992. More than one listing may appear. Click the information icon **(i)** to the right for all entries, then grab the right-side double line and click and drag to the right to expand the text box as far as it will go. You will note that the listings are from different databases. Starry Night has this object appearing in multiple (different) databases (or lists). All refer to the same object; however, the information given may be slightly different, and the coordinates may vary slightly as well. Double-click on the listing associated with the **Bright NGC Objects** database. Adjust the **zoom controls** as needed. Right click (control click for Mac) on the object and select **info**, click the plus sign (gray arrow for Mac) to expand all layers, then complete the following table of information.

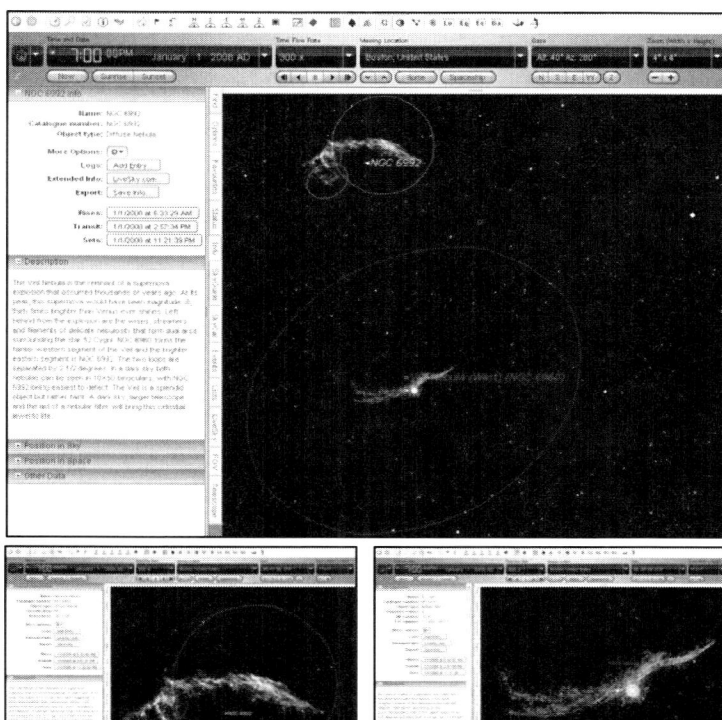

Name and catalog number:	NGC 6992	NGC 6960
Date, time:		
Azimuth, altitude:		
Distance from observer (give units):		
Apparent magnitude:		
Angular size (give units—degrees, arc minutes, etc.):		

What do you find most interesting? Indicate some additional info about these objects:

PART 3: LOTS AND LOTS OF SUPERNOVA REMNANTS

To get a better sense of the number of supernova remnants and their spatial distribution in our sky, you may turn on and label additional supernova remnants by clicking the box to the left of the **supernova remnants database** listing in the **other** layer of the OPTIONS **side pane**. Also click the **label** box to the right.

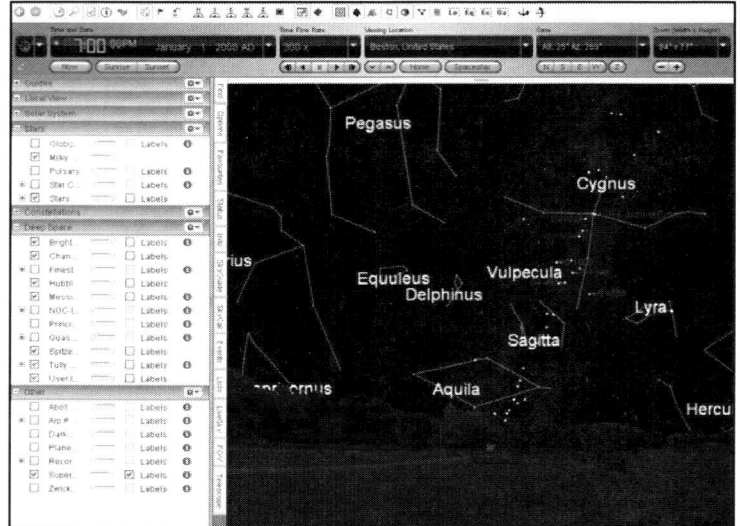

Since pulsars are also produced by certain types of supernovae, you can turn pulsars on by checking the box to the left of **pulsars** in the **stars** layer of the OPTIONS **side pane**, and turn their labels on as well by checking the box to the right.

A small, bright marker will now identify suspected supernova remnants on your screen. A slightly purple marker will identify pulsars. Since the supernova remnant markers look like bright stars, it's helpful to turn off the stars by unchecking the box to the left of **stars** in the **stars** layer of the OPTIONS **side pane**. You should also uncheck the **labels** boxes for **bright NGC objects** and **Messier objects** after you click the plus sign (gray arrow for Mac) to expand the **deep space** layer of the OPTIONS **side pane**.

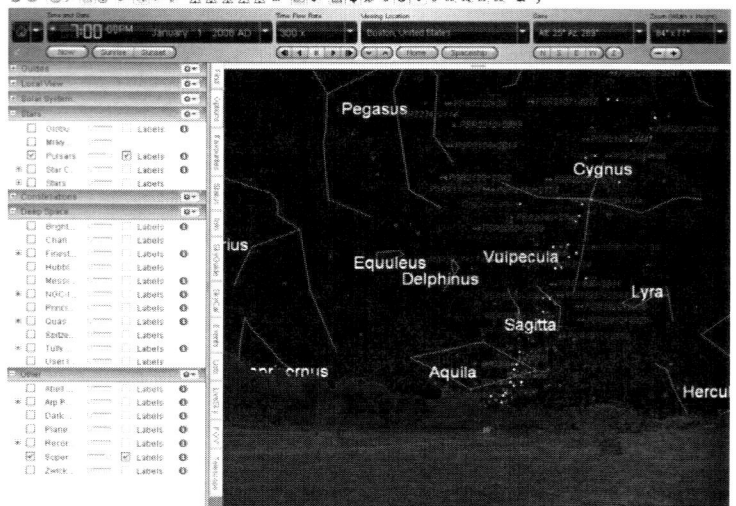

You should now be able to clearly see the distribution of supernova remnants and pulsars, without interference from stars, NGC objects, and Messier objects. The constellation outlines should still be on your screen, though, for reference.

How many supernova remnants are out there? Are there more? Astronomers have only been able to view a small portion of our galaxy. Most of our galaxy is hidden behind dust in the plane of our Milky Way. In fact, we are unable to even see to the center of our own galaxy without the aid of infrared and radio telescopes (longer wavelength radiation passes more easily through interstellar dust). So, of course, the Starry Night databases include only supernova remnants that have been discovered, recorded, and identified as such. In addition, the Starry Night program only shows objects that you would reasonably be able to see with a small telescope. If you zoom in, however, you simulate viewing conditions through a larger telescope and the program will adjust, showing and labeling additional objects. To get a better sense of spatial distribution, although we can't do anything about the supernova remnants that are hidden from view and have never been discovered, we can adjust settings to show us all the objects that have been discovered and included in our database.

Click the OPTIONS tab on the left **side pane**. Click the plus sign (gray arrow for Mac) to expand the **stars**, **deep space**, and **other** layers. Check to be sure only the **supernova remnants** and **pulsars** databases are turned on (check the check box to the left). Also be sure the **labels** box on the right is checked as well. You may also click the information icon **(i)** to the right to read about each database.

Now hover your mouse pointer over the words **supernova remnants** and click to adjust display options. Slide the **number of objects** bar and the **label** bar all the way to the right to maximize the number of objects shown and labeled. Do the same with the **pulsars** database.

Once you have displayed as many supernova remnants and pulsars as you can, use the hand tool to move across the sky to get a sense of the spatial distribution of these objects. Do you notice anything? Do they seem to be concentrated in some areas more than others? Turn on the various coordinate systems by clicking the **coordinate system** buttons* along the **button bar** and see if you can determine what the general alignment is. Note that there are two different concentration areas. Can you pose a possible explanation for this? Why aren't supernova remnants evenly distributed throughout our sky? Why is the distribution of supernova remnants different than the distribution of pulsars? Think about any selection effects that might be involved. For example, a supernova shockwave can excite inert gases and cause them to glow. Where might you expect to find most of the gas and dust in our galaxy? Does this correspond to the distributions you are observing?

*The **lo** button displays a *local coordinate system* based on *azimuth* and *altitude*. The **eq** button displays an *equatorial coordinate system* based on *right ascension* and *declination* (it's similar to Earth's longitude and latitude). The **ec** button displays an *ecliptic coordinate system* keyed to the *ecliptic plane* (the Earth's orbit around the Sun). The **ga** button displays a *galactic coordinate system* keyed to the plane of our supernova remnants.

Activity 27—HERTZSPRUNG–RUSSELL DIAGRAM (H–R DIAGRAM)

*These activities are designed to work with the Starry Night software that comes with your text, from any home location you choose, and with the current date and time, unless indicated otherwise. You may always revert to factory default settings by clicking **FILE/ preferences**, then selecting **factory defaults** as needed. You may also undo a command or series of commands on the PC by clicking the **back** button at the top left of the **button bar**. You should refer to the key given at the beginning of this booklet for clarification of "on screen" buttons, controls, and functions. PC **button bar** items can all be accessed through the **menu**. "Right click" on the PC is equivalent to "control click" on the Mac. All activities assume that OpenGL graphics capabilities are enabled on your computer.*

PART 1: THE HERTZSPRUNG–RUSSELL DIAGRAM (H–R DIAGRAM)

The H–R diagram is a plot of luminosity vs. temperature. Every star in the sky has its own specific absolute brightness and its own characteristic surface temperature. If we were to plot this brightness (on the vertical axis) as a function of the temperature (on the horizontal axis), we would see a general trend from the upper left (hot, bright, and blue stars) to the lower right (cool, dim, and red stars). Note that temperature is related to color, with hot stars appearing bluish and cool stars appearing reddish. Stars plotted on the H–R diagram from upper left to lower right are referred to as "main sequence" stars. These stars are all burning hydrogen fuel and converting it to helium in nuclear fusion reactions at the star's core. We also find groupings of stars in the upper right and lower left parts of the diagram. The stars in the upper right have initiated higher stages of nuclear fusion burning, and represent stars at advanced stages of their stellar evolutionary life cycle. Stars at the lower left of the diagram are white dwarfs, and represent the hot, dense remains of small to medium-sized stars.

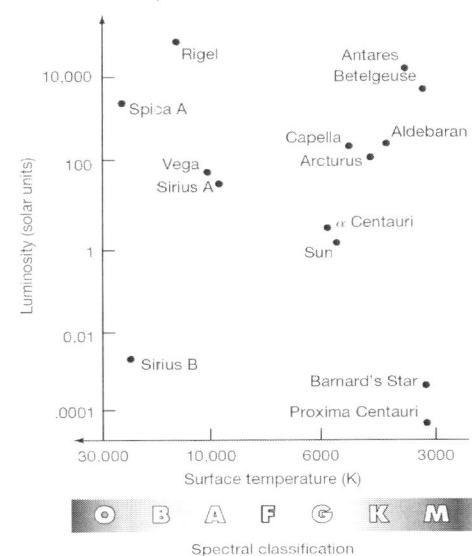

The H–R diagram helps us to visually follow a star's stellar evolutionary life cycle. Stars spend most of their lives on the main sequence fusing hydrogen to helium. After they have used up their hydrogen fuel, low-mass stars turn into red giants, then form planetary nebulae with hot, dense cores called white dwarfs. Higher-mass stars turn into red giants, then move off the main sequence toward the upper right of the H–R diagram as they initiate higher-stage nuclear fusion reactions. Such stars may end their lives in spectacular supernovae, sometimes forming neutron stars, pulsars, or black holes. Thus large giants and supergiants are found in the upper right of the diagram, and small white dwarfs are found in the lower left of the diagram. On the main sequence, massive, hot, blue, bright, and short-lived stars lie at the upper left. Low-mass, cool, red, dim, and long-lived stars lie at the lower right.

PART 2: STARS OF OUR CURRENT EVENING SKY

You should begin this activity at sunset. An easy way to do this is to click the drop-down arrow to the right of the **date & time** field on the **control panel**, and select **sunset**. Look toward the west by clicking the **W viewing direction** button located on the **button bar** across the top of your screen, or by simply keying in the letter W (Mac users should refer to the button bar commands given at the beginning of

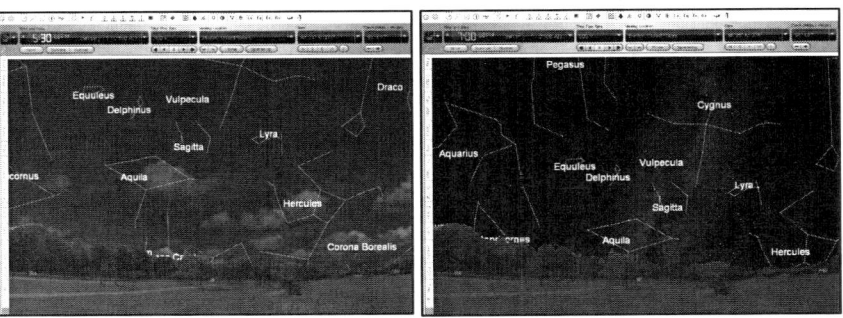

this booklet). The screen will pan toward the west. Select a playing speed of **300×** normal time by clicking the drop-down menu at the right of the **time speed** field. Click the stop **time mode** button when the Sun has set, the stars have come out, and dusk is almost over. Then click on the **constellations** button to show the constellations, and click the **labels** button to label stars and objects.

[OPTIONAL] If you wish to turn other objects "off" that may distract from our survey of stars, click the plus sign (gray arrow for Mac) to expand the **solar system** layer of the OPTIONS **side pane** and turn off items such as **satellites**, **planets-moons**, **comets**, **asteroids**, etc.

[OPTIONAL] You can increase the number of stars being labeled on your screen by Starry Night by expanding the **stars** layer of the OPTIONS **side pane**, then hover your pointer over the word **stars**, then click to change viewing options. Slide the **labels** bar all the way to the right, then click **OK**. Don't slide the other bars to increase the number of stars shown and/or labeled (at least not yet). Starry Night has such a large database that doing this can dramatically decrease your system's performance.

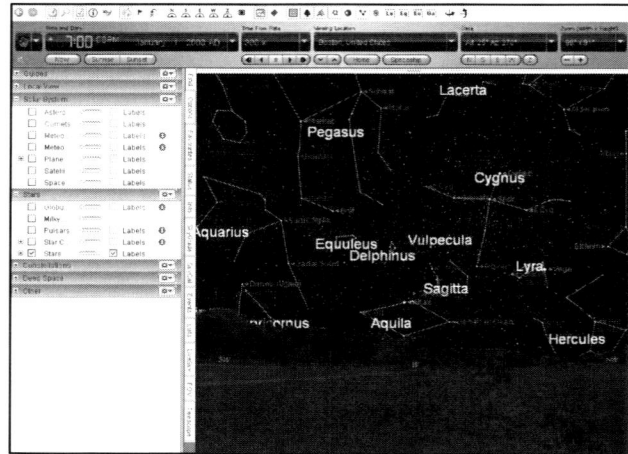

PART 3: PLOTTING STARS ON THE H–R DIAGRAM

Let's examine four characteristically different stars to see where their positions fall on the H–R diagram. We will view each star's location on the H–R diagram as compared to other stars in our night sky. Instead of plotting the luminosity vs. temperature yourself from the raw data (provided by Starry Night), we can make use of Starry Night H–R diagram feature to plot the information for us. We will select a few representative examples, a typical field star such as Alpha Centauri (one of our nearest neighbors and about the same size as our Sun), a red giant/supergiant such as Betelgeuse (in Orion, the mighty hunter), a blue giant such as Deneb (in Cygnus, the swan), and Barnard's star (a small, dim, nearby star that can't even be seen with the naked eye).

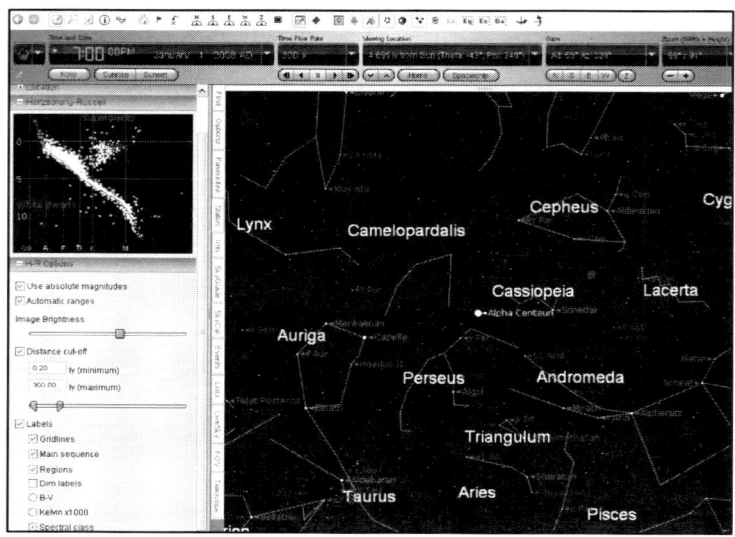

Click the FIND tab on the left **side pane**. In the search field at the top, type in ALPHA CENTAURI. Right click (control click for Mac) on the object and select **show info**. At the top of the INFO **side pane**, click the button labeled **status**. Click the plus sign (gray arrow for Mac) to the left of **Hertzsprung–Russell** to expand this layer. Also click the plus sign (gray arrow for Mac) to the left of **H–R options** to expand this layer as well. You can use this layer to adjust the plotting options. You may need to wait a bit for Starry Night to access its database and populate the H–R plot. The graph you see has plotted all the stars shown on the screen. If you move across the sky and/or zoom in or out, the graph will update in real time, adjusting for what stars are shown on your screen at any given moment. Zoom out all the way and hover your mouse pointer over Alpha Centauri (or any star in your field of view). You should see a red, flashing light at this star's plotted point on the H–R diagram.*

*If you run into any problems, repeat the full procedure described here, starting with the FIND tab on the left **side pane**. This can help reinitialize the plotting function.

Now use this procedure to plot on your own hand drawn H–R diagram of the positions of each of the following stars. *(For more information on a specific star, refer back to its individual activity.)*

- ALPHA CENTAURI (HIP71683)
- BETELGEUSE (HIP27989)
- DENEB (HIP102098)
- BARNARD'S STAR (HIP71683)

You should outline the main sequence on your plot and be sure to show the correct relative position of your plotted points to the main sequence. Be sure to include values for the axes. Be careful to place the star positions correctly so that they correspond to the correct axis values for temperature and luminosity.* All four stars should be plotted on the same graph. **

*Label luminosity on the vertical axes and temperature on the horizontal axis, and plot higher temperatures to the left and lower temperatures to the right. You may also express the vertical axes as absolute magnitude and the horizontal axes as stellar spectral type, B-V, or color (blue to red).

**If you are curious as to where our Sun would fall on this plot, it would not be far off from Alpha Centauri (perhaps a little bit to the lower right on the main sequence).

PART 4: DISCUSSION OF STELLAR CHARACTERISTICS IN RELATION TO H–R DIAGRAM

The four stars you plotted are characteristically different. Alpha Centauri (one of our nearest neighbors and about the same size as our Sun) is a typical field star, Betelgeuse (in Orion, the mighty hunter) is a red giant/supergiant, Deneb (in Cygnus, the swan) is a blue giant/supergiant, and Barnard's star (a small, dim, nearby star that can't even be seen with the naked eye) is a red dwarf. Let's learn more about each of these types of stars. You can use the Starry Night software to search for these as we discuss them. Right click (control click for Mac) and select **show info** for more details.

ALPHA CENTAURI (HIP71683)

Alpha Centauri is a typical star on the main sequence. Note how this star's radius, temperature, and luminosity compare to our Sun. Such stars are fusing hydrogen to helium. This is a most efficient nuclear fuel (0.7% conversion to pure energy by Einstein's $E = mc^2$). Stars will remain on the main sequence for millions, billions, or even trillions of years. A star's mass determines how long its fuel will last. Massive stars burn fuel faster due to greater gravitational forces in their core, producing higher pressures and temperatures. Such stars may only last millions of years before consuming their fuel. This may seem paradoxical since large-mass stars have more fuel to burn. However, their greater mass does not compensate for the even greater rate of nuclear fusion. A star does not move much from its position on the main sequence throughout its hydrogen-burning phase (a slight brightening does occur, but this results in only a slight shift off the main sequence toward the upper right).

BETELGEUSE (HIP27989)

Betelgeuse is a typical red giant/supergiant. These types of stars are near the end of their stellar lives. Having run out of hydrogen fuel, this star has left the main sequence to become brighter, yet cooler and redder. It has moved toward the upper right of the H–R diagram. The star's core contracts, further heating the central core, yet expanding its outer shell. Depending on their mass, red giants/supergiants may end up as planetary nebulae (with a central white dwarf) if $M_{core} < 1.4$ solar masses; they may go supernova, resulting in pulsars (spinning neutron stars) if $M_{core} > 1.4$ solar masses, or even become black holes (gravitationally collapsed objects whose core material after going supernova must be greater than 3 solar masses).

DENEB (HIP102098)

Deneb is a blue giant/supergiant. Note the surface temperature of this star as compared to our Sun. Hotter temperatures imply a bluish color. Cooler temperatures imply a reddish color. Also note the radius, luminosity, and distance of this star. This star is surprisingly bright for being so very far away from us. Such tremendous power output is due to the fourth power relationship of brightness to temperature. Higher temperatures have a greater effect on brightness than do size and distance, which vary only by the square of the variable.

BARNARD'S STAR (HIP71683)

Barnard's Star is a low-mass star that is very dim. Note how this star's mass, radius, temperature, and luminosity compare to our Sun. This star is barely massive enough to have initiated nuclear fusion. Although this star is relatively close to us, it is all but impossible to see without the aid of high-powered telescopes and equipment. Compare this star's luminosity characteristics to that of Deneb discussed earlier. Such comparisons can serve to give us a greater appreciation of the full extent of luminosity ranges that stars can have. Another interesting and somewhat paradoxical consideration is to realize that the stars we see at night are not a representative sample of the stars in our galaxy. Although Barnard's star is one of the closest stars to us, we can't even hope to see it with the naked eye. However, Deneb (along with other giants/supergiants) appears as one of the brightest stars in the sky, even at thousands of light-years' distance. This biases our perspective of the heavens. We *think* stars are bluish-whitish. However, most stars are orangish–reddish, but just too dim to see. We think stars are bright; however, most stars are relatively dim, so dim in fact that we can't even see most of them. So the typical star is not as we see it in the night sky. Careful surveys are needed to get a sense of the true distribution of stellar characteristics, as seen or not seen by the naked eye.

PART 5: [OPTIONAL] LUMINOSITY DISCUSSION & CALCULATIONS

What does the magnitude and luminosity information you found for each of these stars mean? The primary factors that affect a star's brightness are distance, temperature, and size.

Brightness varies by the inverse square of distance. Thus, a star twice as far away is one-quarter $(1/2^2)$ the brightness. A star three times as far away is one-ninth $(1/3^2)$ the brightness.

Brightness varies by the fourth power of temperature, where hot stars are brighter than dim stars. A star twice as hot is 16 (2^4) times brighter. A star three times as hot is 81 (3^4) times brighter.

Brightness also varies by the square of the radius. Although the star is a sphere, only the cross-sectional solid angle contributes to the brightness we see. Thus a star twice as big is four (2^2) times as bright. A star three times as big is nine (3^2) times as bright.

Use these relationships to determine how much brighter (or dimmer) a hypothetical star with the given characteristics would be as compared to Alpha Centauri, a sun-like star a little over 4 light-years away. These are "order of magnitude" calculations (to keep the math simple). For simplicity, state whether the configuration described for the star at the left is *brighter* or *dimmer* than for Alpha Centauri. Alternatively, you may express your answer numerically.

Characteristics of Hypothetical Star *(to be compared to Alpha Centauri)*	**Multiple of Alpha Centauri's brightness** *(expressed qualitatively or quantitatively)*
$2 \times$ Distance, $1 \times$ Temperature, $2 \times$ Size:	
$2 \times$ Distance, $2 \times$ Temperature, $2 \times$ Size:	
$100 \times$ Distance, $3 \times$ Temperature, $10 \times$ Size (Blue Giant Star):	
$100 \times$ Distance, $4 \times$ Temperature, $10 \times$ Size (Blue Supergiant Star):	
$100 \times$ Distance, $(\frac{1}{2}) \times$ Temperature, $100 \times$ Size (Red Supergiant Star):	
$100 \times$ Distance, $10 \times$ Temperature, $(1/100) \times$ Size (White Dwarf Star):	

Activity 28—GALAXIES

*These activities are designed to work with the Starry Night software that comes with your text, from any home location you choose, and with the current date and time, unless indicated otherwise. You may always revert to factory default settings by clicking **FILE/ preferences**, then selecting **factory defaults** as needed. You may also undo a command or series of commands on the PC by clicking the **back** button at the top left of the **button bar**. You should refer to the key given at the beginning of this booklet for clarification of "on screen" buttons, controls, and functions. PC **button bar** items can all be accessed through the **menu**. "Right click" on the PC is equivalent to "control click" on the Mac. All activities assume that OpenGL graphics capabilities are enabled on your computer.*

PART 1: NEBULAE

The term *nebulae* initially included all "fuzzy" objects seen in telescopes, but now many of them have been identified as galaxies, open clusters, globular clusters, stellar associations, supernova remnants, planetary nebulae, emission nebulae, reflection nebulae, and dark nebulae. This activity will focus on galaxies.

Galaxies are large collections of stars numbering in the millions, billions, or trillions, revolving around a common galactic center that most likely includes a supermassive black hole. Galaxies can have many shapes, including elliptical and spiral (such as our own). Galaxies are often millions of light-years apart from each other with essentially nothing in between.

PART 2: FINDING GALAXIES IN OUR NIGHT SKY

You should begin this activity at sunset. An easy way to do this is to click the drop-down menu to the right of the **date & time** field on the **control panel**, and select **sunset**. Look toward the west by clicking the **W viewing direction** button located on the **button bar** across the top of your screen, or by simply keying in the letter W (Mac users should refer to the button bar commands given at the beginning of this booklet). The screen will pan toward the west. Select a playing speed of **300×** normal time by clicking the drop-down menu at the right of the **time speed** field. Click the STOP **time mode** button when the Sun has set, the stars have come out, and dusk is almost over. Then click on the **constellations** button to show the constellations.

Click the OPTIONS tab on the left **side pane** and expand the **deep space** layer by clicking on the small plus sign to the left (gray arrow for Mac). Activate the labels for both the **bright NGC objects** and **Messier objects** by clicking the **labels** box to the right. Then click the information icon **(i)** to the far right to display descriptive information on these Starry Night databases.

You should now see numerous deep-space objects identified on your screen.

[OPTIONAL] For a more complete listing of galaxies, you may turn on and label additional galaxies by clicking the box to the left of the **NGC-IC database** listing in the **other** layer of the OPTIONS side pane. Don't click the **label** box to the right. Click the plus sign to the left (gray arrow for Mac), then select **galaxy** by checking the box to the left while unchecking all the others. Again, don't click the **label** box to the right. A small marker will now identify additional galaxies on your screen. For this activity, scan the sky only for those that are also labeled since only these will display an image in Starry Night. Take careful note of the fact that this is an overlapping database. This means that some of the objects in this database are already displayed in Starry Night by other databases. This can lead to confusion since the coordinates may not be exactly the same, thus appearing to indicate that the same object is in two slightly different places at the same time. Although not exact, the additional database markers can help us locate objects. As you zoom in, you should ignore the marker and look in the general vicinity for the intended object. Once you locate it, right click (control click for Mac) and center it on your screen and continue zooming in. Be careful not to confuse galaxies with other **Messier** and **NGC** objects. The on screen info will confirm the **object type** when you hover your mouse over it.

PART 3: EXAMPLES OF GALAXIES

Let's look at some well-known examples. Astronomers always have a few favorites. Which of these is your favorite?

THE ANDROMEDA GALAXY (M31):

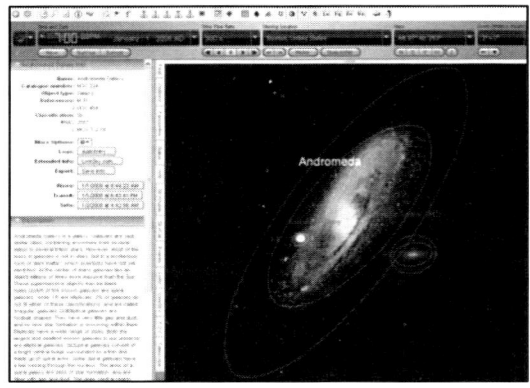

Click the FIND tab on the left **side pane**. In the search field at the top, type in M31. More than one listing may appear. Click the information icon **(i)** to the right for all entries, then grab the right-side double line and click and drag to the right to expand the text box as far as it will go. You will note that the listings are from different databases. This simply means that Starry Night has this object appearing in multiple (different) databases (or lists). All refer to the same object; however, the information given may be slightly different, and the coordinates may vary slightly as well. Double-click on the listing associated with the **Messier object** database. Zoom in on the object using the **zoom controls** at the right of the **control panel**. Right click (control click for Mac) on the object and select **show info**, click the plus sign (gray arrow for Mac) to expand all layers, and then complete the following table of information.

Date, time:	
Azimuth, altitude:	
Distance from observer (give units):	
Apparent magnitude:	
Angular size (give units—degrees, arc minutes, etc.):	

What do you find most interesting? Indicate some additional info about this object:

THE SOMBRERO GALAXY (M104):

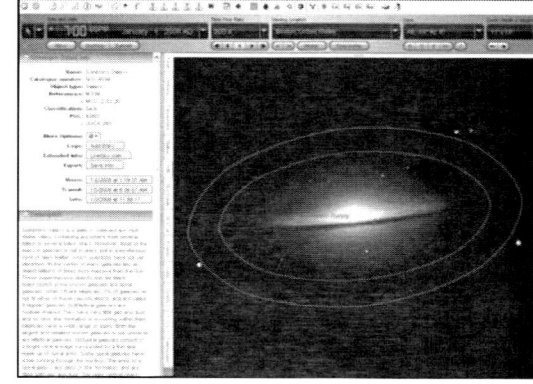

Click the FIND tab on the left **side pane**. In the search field at the top, type in M104. More than one listing may appear. Click the information icon **(i)** to the right for all entries, then grab the right-side double line and click and drag to the right to expand the text box as far as it will go. You will note that the listings are from different databases. Starry Night has this object appearing in multiple (different) databases (or lists). All refer to the same object; however, the information given may be slightly different, and the coordinates may vary slightly as well. Double-click on the listing associated with the **Messier objects** database. Adjust the **zoom controls** as needed. Right click (control click for Mac) on the object and select **show info**, click the plus sign (gray arrow for Mac) to expand all layers, then complete the following table of information.

Date, time:	
Azimuth, altitude:	
Distance from observer (give units):	
Apparent magnitude:	
Angular size (give units—degrees, arc minutes, etc.):	

What do you find most interesting? Indicate some additional info about this object:

THE WHIRLPOOL GALAXY (M51):

Click the FIND tab on the left **side pane**. In the search field at the top, type in M51. More than one listing may appear. Click the information icon **(i)** to the right for all entries, then grab the right-side double line and click and drag to the right to expand the text box as far as it will go. You will note that the listings are from different databases. Starry Night has this object appearing in multiple (different) databases (or lists). All refer to the same object; however, the information given may be slightly different, and the coordinates may vary slightly as well. Double click on the listing associated with the **Messier objects** database. Adjust the **zoom controls** as needed. Right click (control click for Mac) on the object and select **show info**, click the plus sign (gray arrow for Mac) to expand all layers, and then complete the following table of information.

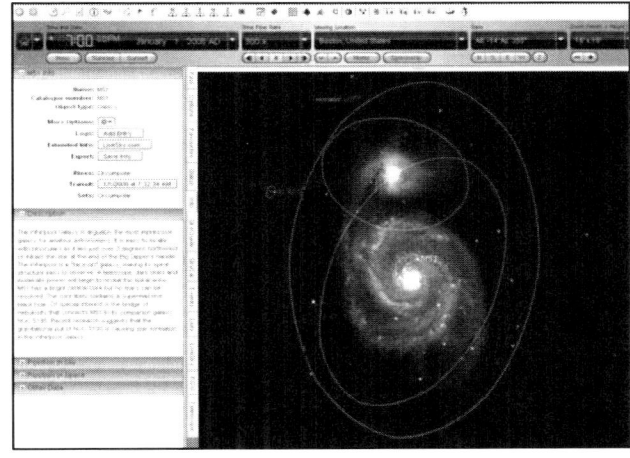

Date, time:	
Azimuth, altitude:	
Distance from observer (give units):	
Apparent magnitude:	
Angular size (give units—degrees, arc minutes, etc.):	

What do you find most interesting? Indicate some additional info about this object:

PART 4: FINDING YOUR OWN EXAMPLES

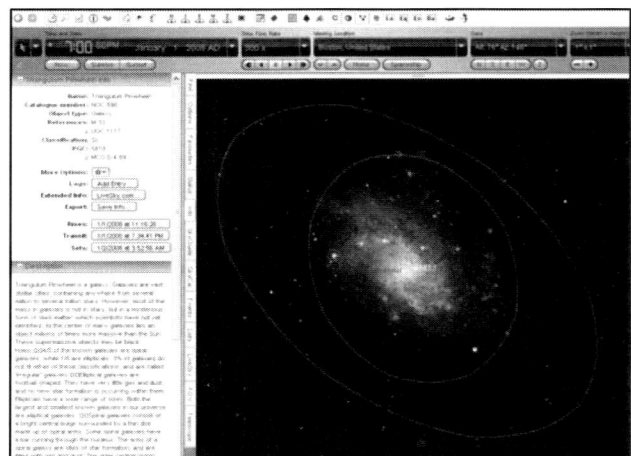

In this section, we are interested in finding our own examples of galaxies. Zoom all the way out, return to just after sunset by following the instructions given in Part 2, and scan your current evening sky for a galaxy. You may do this either by using the **viewing direction** buttons N, S, E, W, and Z on the **button bar**, or by simply "grabbing" the sky with the **hand tool** and dragging while holding the left mouse button down. You may need to zoom in and out a bit to see the dimmer objects. You may also advance time forward to later in the evening if you wish. Read the descriptions given for each object until you find one identifying it as a galaxy. You can confirm that the object is a galaxy by reading what the **object type** is when hovering your mouse over it. Right click (control click for Mac) on this object (careful to hover the mouse exactly over the object as best you can) and select **centre**. Once centered, zoom in by using the **zoom control** at the right side of the **control panel**. If your chosen object begins to drift from the center of the screen, stop the zooming process and right click (control click for Mac) once again on the object, selecting **centre**.

Once you have zoomed in, right click (control click for Mac) on the object and select **show info** to learn more about it. Click the plus sign (gray arrow for Mac) to expand all layers.

Galaxy #1: Use the information provided by Starry Night to complete the following table:
(Try not to select any of the same examples already given in this activity.)

Date, time:	
Azimuth, altitude:	
Name and catalog number:	
Distance from observer (give units):	
Apparent magnitude:	
Angular size (give units—degrees, arc minutes, etc.):	

Provide a sketch of the object as you see it in Starry Night:

Galaxy #2: Use the information provided by Starry Night to complete the following table:
(Try not to select any of the same examples already given in this activity.)

Date, time:	
Azimuth, altitude:	
Name and catalog number:	
Distance from observer (give units):	
Apparent magnitude:	
Angular size (give units—degrees, arc minutes, etc.):	

Provide a sketch of the object as you see it in Starry Night:

Galaxy #3: Use the information provided by Starry Night to complete the following table:
(Try not to select any of the same examples already given in this activity.)

Date, time:	
Azimuth, altitude:	
Name and catalogue number:	
Distance from observer (give units):	
Apparent magnitude:	
Angular size (give units—degrees, arc minutes, etc.):	

Provide a sketch of the object as you see it in Starry Night:

PART 5: LOTS AND LOTS OF GALAXIES

How many galaxies are out there? Astronomers cannot see in all directions clearly. The dust in the plane of our own Milky Way galaxy obscures our view. Luckily, we are able to see many galaxies by looking above and below our galactic plane. The Starry Night databases include only galaxies that have been discovered, recorded, and cataloged. In addition, the Starry Night program only shows objects that you would reasonably be able to see with a small telescope. If you zoom in, however, you simulate viewing conditions through a larger telescope and the program will adjust, showing and labeling additional objects. To get a better sense of spatial distribution, although we can't do anything about the galaxies that are hidden from view and have never been discovered, we can adjust settings to show us all the objects that have been discovered and included in our database.

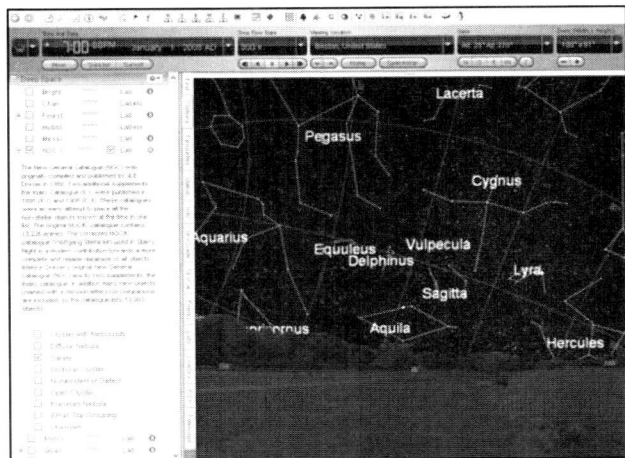

Click the OPTIONS tab on the left **side pane**. Click the plus sign (gray arrow for Mac) to expand the **deep space** and **other** layers. Turn off the **bright NGC objects** and **Messier objects** databases, then click the plus sign (gray arrow for Mac) to expand the **NGC-IC** database and make sure that only the **galaxy** check box is turned on. Click the information icon **(i)** to the right to read about this database. Select the check box on the left and click the **label** box on the right. Now hover your mouse pointer over the words **NGC-IC** and click to adjust display options. Slide the **number of objects** bar and the **label** bar all the way to the right to maximize the number of objects shown and labeled.

Once you have displayed as many galaxies as you can, use the hand tool to move across the sky to get a sense of the spatial distribution of these objects. Do you notice anything? Do they seem to be concentrated in some areas more than others? Turn on the various coordinate systems by clicking the **coordinate system** buttons* along the **button bar** and see if you can determine the general alignment. Note that there are two different concentration areas. Can you pose a possible explanation for this? Why don't galaxies appear to be evenly distributed throughout our sky?

*The **lo** button displays a *local coordinate system* based on *azimuth* and *altitude*. The **eq** button displays an *equatorial coordinate system* based on *right ascension* and *declination* (it's similar to Earth's longitude and latitude). The **ec** button displays an *ecliptic coordinate system* keyed to the *ecliptic plane* (the Earth's orbit around the Sun). The **ga** button displays a *galactic coordinate system* keyed to the plane of our galaxy.

Activity 29—VIRGO GALAXY CLUSTER

*These activities are designed to work with the Starry Night software that comes with your text, from any home location you choose, and with the current date and time, unless indicated otherwise. You may always revert to factory default settings by clicking **FILE/ preferences**, then selecting **factory defaults** as needed. You may also undo a command or series of commands on the PC by clicking the **back** button at the top left of the **button bar**. You should refer to the key given at the beginning of this booklet for clarification of "on screen" buttons, controls, and functions. PC **button bar** items can all be accessed through the **menu**. "Right click" on the PC is equivalent to "control click" on the Mac. All activities assume that OpenGL graphics capabilities are enabled on your computer.*

PART 1: FINDING THE VIRGO GALAXY CLUSTER IN THE SKY

The Virgo cluster is a large cluster of galaxies in a direction bordering the constellations Virgo and Leo. This activity will explore this region of the sky as seen from Earth, and as viewed from within the cluster itself.

You should begin this activity at sunset. An easy way to do this is to click the drop-down menu to the right of the **date & time** field on the **control panel**, and select **sunset**. Look toward the west by clicking the **W viewing direction** button located on the **button bar** across the top of your screen, or by simply keying in the letter W (Mac users should refer to the button bar commands given at the beginning of this booklet). The screen will pan toward the west. Select a playing speed of **300×** normal time by clicking the drop-down menu at the right of the **time speed** field. Click the STOP **time mode** button when the sun has set, the stars have come out, and dusk is almost over. Then click on the **constellations** button to show the constellations.

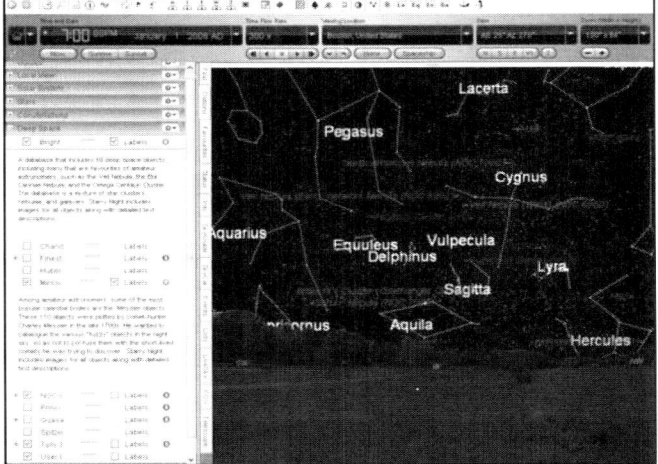

Click the OPTIONS tab on the left **side pane** and expand the **deep space** layer by clicking on the small plus sign to the left (gray arrow for Mac). Activate the labels for both the **bright NGC objects** and **Messier objects** by clicking the **labels** box to the right. Then click the information icon **(i)** to the far right to display descriptive information on these Starry Night databases.

You should now see numerous deep-space objects identified on your screen. These are some of the most famous telescope viewing objects in the sky.

To find the Virgo constellation, click the FIND tab on the left **side pane** at the left of your screen. Type in the name of the constellation you are trying to find. Type the name in slowly and a list of options will be shown. Don't type out the entire name of the constellation, or you will get an expanded list of all the stars in the constellation. Three or four letters should suffice. From the list shown, double-click the item listed as a CONSTELLATION in the second column (the column that indicates the kind of object being shown). This will center the constellation on your screen and fully illustrate it for you.

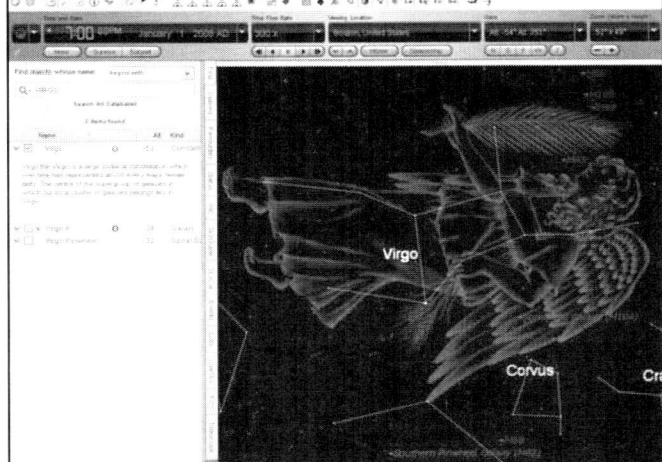

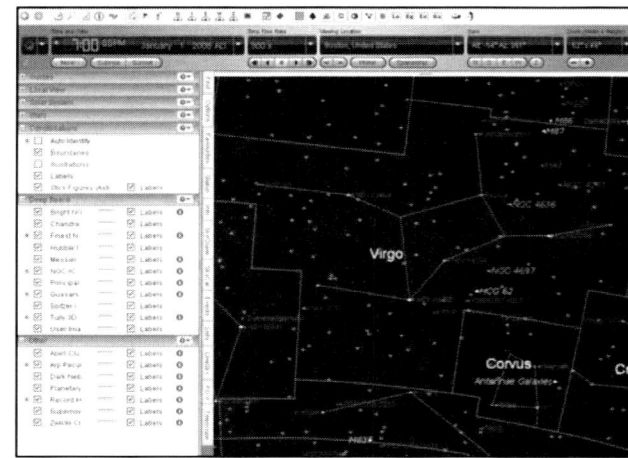

If the constellation is below the horizon, you may need to select a new time using the **date & time** controls on the **control panel**, and/or turn off the Earth's horizon by clicking the **horizon** button on the **button bar**. If the constellation is up only in the daytime, then you will also need to turn off the Sun by clicking the **daylight** button on the **button bar**. To learn more about this constellation, click the information icon (**i**) to the right of its listing in the FIND **side pane**.

Right click (control click for Mac) on the constellation itself as shown on your screen (careful to not hover over a star or object while doing this) and **deselect** to turn off the full illustration. Click the **galaxies** button on the **button bar** (this turns on the **NGC-IC da**tabase). You should now see only the stick-figure representation of the constellation with little circles and colored dots to indicate deep space objects of observational interest.

Click the VIEW OPTIONS tab on the left **side pane** and click the plus sign (gray arrow for Mac) to expand the **constellations** layer. Click the **boundaries** check box to outline constellation boundaries. Now zoom in until the constellation almost fills your screen. The Virgo Cluster is the large group of galaxies between Virgo and Leo.

[OPTIONAL] You can increase the number of objects being labeled by Starry Night by expanding the **deep space** and **other** layers of the OPTIONS **side pane**, then hovering your pointer over the words **bright NGC objects**, then clicking to change viewing options. Slide the **labels** bar all the way to the right, then click **OK**. Do the same with **Messier objects**. Follow the same procedure for the **NGC-IC database**, however, slide the **show** bar to the right (don't worry about the **labels** bar in this case since it will be turned off). This will increase the number of objects shown and labeled on your screen. Although the **bright NGC objects** and **Messier objects** should be labeled (right check box), be sure not to select the labeling option for the **NGC-IC** database

PART 2: THE VIRGO CLUSTER AS VIEWED FROM *HERE*

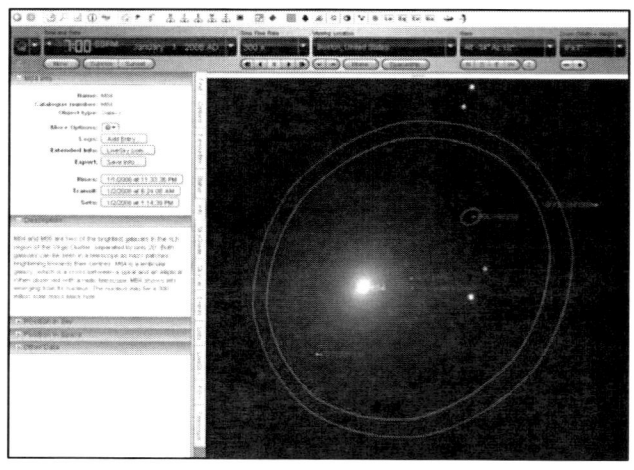

Look for M84 labeled within the boundaries of the Virgo constellation as shown on your screen.* If an object is not labeled on your screen, hovering your mouse pointer over the object will give you some identifying information. Once you locate what you are looking for, zoom in on the object using the **zoom controls** at the right of the **control panel**. Right click (control click for Mac) on the object and select **show info**, click the plus sign (gray arrow for Mac) to expand all layers.

M84, M86, M87, M58, M89, M90, M60, and M59 all lie in the direction of the Virgo cluster. Many other galaxies of the Virgo cluster can be found next to these along the border of neighboring constellation Leo. Zoom in on M84, M86, and M87, then use the information provided by Starry Night to complete the following table for each object viewed:

Name and catalog number:	**M84**	**M86**	**M87**
Date, time:			
Azimuth, altitude:			
Distance from observer (give units):			
Apparent magnitude:			
Angular size (give units—degrees, arc minutes, etc.):			

Provide a sketch of your own for these galaxies as you see them in Starry Night:

[ ]

*If you have trouble finding M84, click the FIND tab on the left **side pane**. In the search field at the top, type in M84. More than one listing may appear. Click the information icon **(i)** to the right for all entries to learn more about these objects. Grab the right-side double line, and click and drag to the right to expand the text box as far as it will go. You will note that the listings are from different databases. This simply means that Starry Night has this object appearing in multiple (different) databases (or lists). All refer to the same object; however, the information given may be slightly different, and the coordinates may vary slightly as well. Double-click on the listing associated with the **Messier objects** database if the object name begins with the letter M. Double-click cn the listing associated with the **bright NGC objects** database if the object name begins with the letters NGC. Now zoom in as described above.

PART 3: THE VIRGO CLUSTER AS VIEWED FROM *THERE*

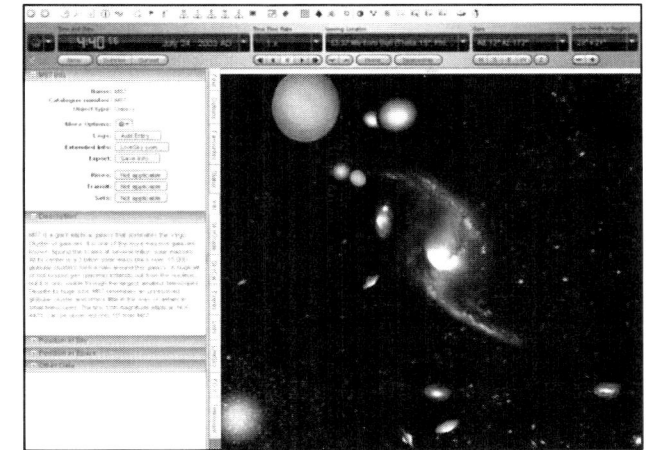

To get a better view of the Virgo cluster, we will travel to a point 56.37 million light-years from home, outside our own galaxy, and looking in the direction of M87. This location has been preset in Starry Night and you can quickly "go" there by selecting the following from the **menu** at the top of your screen: **favourites/deep space/ galaxies/Virgo Cluster**.

To identify M87, click the FIND tab on the left **side pane**. In the search field at the top, type in M87. More than one listing may appear. You can click the information icon **(i)** to the right for all entries to learn more about these objects. Grab the right-side double line, and click and drag to the right to expand the text box as far as it will go. You will note that the listings are from different databases. Starry Night has this object appearing in multiple (different) databases (or lists). All may refer to the same object; however, the information given may be slightly different, and the coordinates may vary slightly as well. Double-click on the listing associated with the tully 3D database. Since we are no longer viewing from within our galaxy, only this database will give correct results. Double-click on the correct listing or right click (control click for Mac) and select **centre**. Now M87 should be shown identified on your screen and appearing as a large dominant galaxy near the center of your field of view.

Let's label the remaining galaxies by clicking on the OPTIONS tab, expanding the **deep space** layer by clicking the plus sign to the left (gray arrow for Mac), then clicking the **label** check box to the right of the **tully 3D** database listing. This will label many of the galaxies on your screen.*

*You can increase the number of galaxies being labeled by Starry Night by hovering your pointer over the words **tully 3D** database, then click to change viewing options. Slide the **labels** bar all the way to the right, then click **OK**.

Now, let's see where M84 and M86 are located. Enter M84 in the FIND tab and double-click on the TULLY 3D DATABASE item from the list shown (or right click and select **centre**).

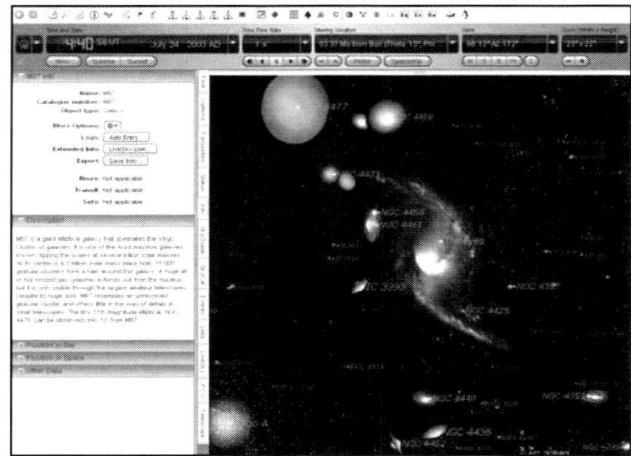

Try looking around in different directions to get a sense of the size and distribution of this large and famous cluster of galaxies. You can do this by "grabbing" and "dragging" the screen (just click and drag anywhere on the viewable area of the screen), or you can use the **viewing direction** buttons N, S, E, W, and Z on the **button bar** (note that these directions are oriented with respect to the galactic coordinate system and have no relation to our local directions of N, S, E, and W). For fun, see if you can find a large spiral galaxy called "The Eyes." Turn off the FIND **side pane** while you do this to increase your viewing area. Just click the FIND **side pane** tab and it will toggle off. To bring the FIND **side pane** back, click the FIND **side pane** tab again.

Finally, to get our bearings, let's look toward our own galaxy. The easiest way to do this is to clear the FIND **side pane** search field. Just erase any letters in the box at the top. Now a list of solar system objects will appear. Double-click on our Sun. The screen will pan to center the location of our own galaxy and solar system. Now zoom in as far as you can using the **zoom control** buttons on the far right of the **control panel**. You will find that we are too far away to see our Sun. Zoom back out before proceeding.

Next, click the down arrow to the left of our **location control** on the **control panel** at the top of the screen. Click and hold this button down and you will see the screen animate a return journey home. Hold the button down continuously and watch the location information in the **location control** area of the **control panel**. You will have to hold the button down for some time before returning to our home solar system. Turn off the FIND **side pane** while you do this to increase your viewing area. Just click the FIND **side pane** tab and it will toggle off.

Illustrate our galaxy as shown in Starry Night as we see and enter it during our return journey. Just draw our galaxy and show the correct orientation from the perspective shown.

Activity 30—VIRGO the MAIDEN (SPRING SKY)

*These activities are designed to work with the Starry Night software that comes with your text, from any home location you choose, and with the current date and time, unless indicated otherwise. You may always revert to factory default settings by clicking **FILE/ preferences**, then selecting **factory defaults** as needed. You may also undo a command or series of commands on the PC by clicking the **back** button at the top left of the **button bar**. You should refer to the key given at the beginning of this booklet for clarification of "on screen" buttons, controls, and functions. PC **button bar** items can all be accessed through the **menu**. "Right click" on the PC is equivalent to "control click" on the Mac. All activities assume that OpenGL graphics capabilities are enabled on your computer.*

PART 1: THE VIRGO CONSTELLATION (SPRING SKY)

In this activity, we are going to explore the constellation Virgo for objects of interest both for users of the Starry Night software and for amateur astronomers using telescopes in the great outdoors. If you don't have a telescope, this is your chance to explore the night sky as if you did. If you do have a telescope, you can survey possible viewing targets for your nighttime observing sessions. Virgo is an evening constellation in the spring sky.

There are many different stories from many different cultures associated with each constellation of the night sky. Western culture is most familiar with those whose origin stem from Greek and Roman mythology. Virgo is the only female figure among the constellations of the Zodiac and may represent a number of different deities.

In one story, Virgo is personified by Ceres, goddess of agriculture, or perhaps her beautiful daughter Proserpina. Pluto, god of the Underworld, taken by Proserpina's tender beauty, abducted her in his chariot and kept her in the dark underworld. Ceres was in such utter despair that she neglected her duties, and so it happened that everything died, nothing would grow, and the Earth was threatened with a worldwide famine. Jupiter could not tolerate this, so he decreed that Pluto should allow Proserpina to be with Ceres in the Upperworld for half the year, and the other half the year she could remain with him in the Underworld. This is why we see the constellation Virgo in the sky from March through August, the season of growing and harvesting, and explains why the land lies barren the rest of the year while Ceres weeps and awaits her daughter's return from the Underworld. As a variant to this theme, the ancient Chaldeans saw the goddess Ishtar in Virgo. Ishtar's husband, Tammuz, was once overpowered by King Winter and dragged off to the Underworld. Ishtar searched for her husband and found him, but became imprisoned in the Underworld as well. The gods forced the keeper of the Underworld to release Ishtar and Tammuz, so that the goddess of agriculture could return to her duties.

In Egypt, the stars of Virgo depicted the goddess Isis. She was once frightened by the monster Typhoon and dropped a sheaf of grain, scattering a trail along the Zodiac. Virgo is sometimes thought to represent Astrea, the goddess of justice. The scales of Libra lie close to Virgo in the sky and was used by Astrea to weigh the good and bad of men.

In this activity, we will take a closer look at the Virgo constellation and examine some of the interesting celestial objects that lie within it.

PART 2: FINDING VIRGO IN THE SKY

You should begin this activity at sunset. An easy way to do this is to click the drop-down menu to the right of the **date & time** field on the **control panel**, and select **sunset**. Look toward the west by clicking the **W viewing direction** button located on the **button bar** across the top of your screen, or by simply keying in the letter W (Mac users should refer to the button bar commands given at the beginning of this booklet). The screen will pan toward the west. Select a playing speed of **300×** normal time by clicking the drop-down menu at the right of the **time speed** field. Click the STOP **time mode** button when the Sun has set, the stars have come out, and dusk is almost over. Then click on the **constellations** button to show the constellations.

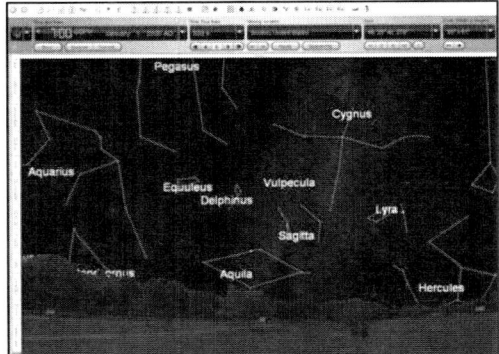

Click the OPTIONS tab on the left **side pane** and expand the **deep space** layer by clicking on the small plus sign to the left (gray arrow for Mac). Activate the labels for both the **bright NGC objects** and **Messier objects** by clicking the **labels** box to the right. Then click the information icon **(i)** to the far right to display descriptive information on these Starry Night databases.

You should now see numerous deep-space objects identified on your screen.

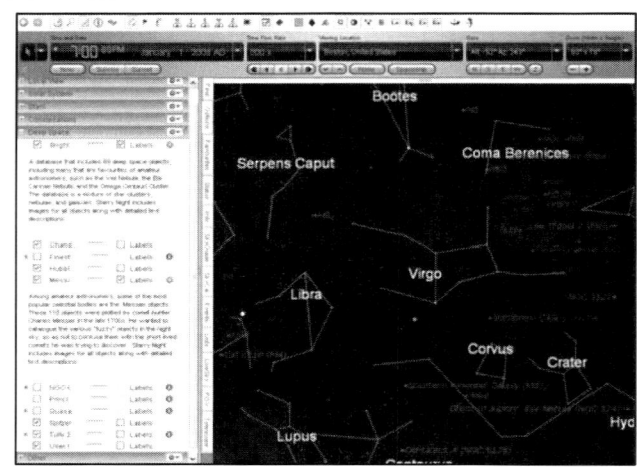

To find the Virgo constellation, click the FIND tab on the left **side pane**. Type in the name of the constellation you are trying to find. (Hint: If you type the name of the constellation slowly a list of options will be offered to you.) Don't type out the entire name of the constellation, or you will get an expanded list of all the stars in the constellation. Three or four letters should suffice. From the list shown, double-click the item listed as a CONSTELLATION in the second column (the column that indicates the kind of object being shown). This will center the constellation on your screen and fully illustrate it for you. If the constellation is below the horizon, you may need to select a new time using the **date & time** controls on the **control panel**, and/or turn off the Earth's horizon by clicking the **horizon** button on the **button bar**. If the constellation is up only in the daytime, then you will also need to turn off the Sun by clicking the **daylight** button on the **button bar**. To learn more about this constellation, click the information icon **(i)** to the right of its listing in the FIND tab of the **side pane**.

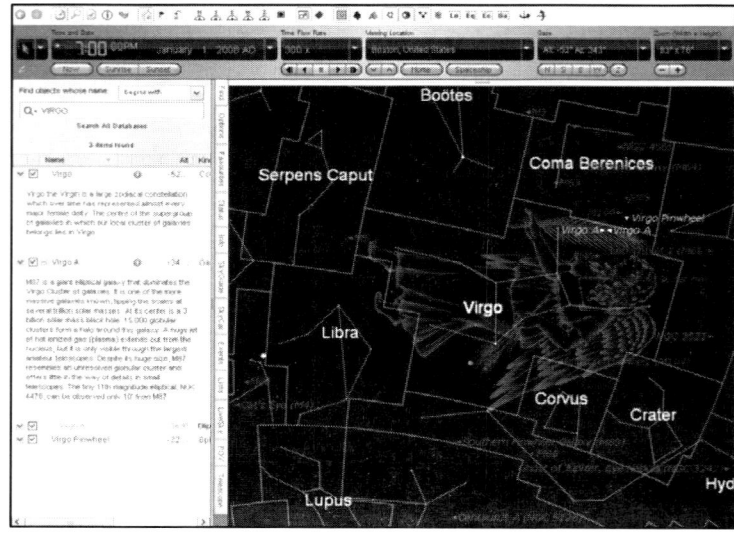

Right click (control click for Mac) on the constellation itself, as shown on your screen (careful not to hover over a star or object while doing this), and **deselect** to turn off the full illustration. Click the **galaxies** button on the **button bar** (this turns on the **NGC-IC** database). You should now see only the stick-figure representation of the constellation with little circles and colored dots to indicate deep-space objects of observational interest.

Click the OPTIONS tab on the left **side pane** and click the plus sign (gray arrow for Mac) to expand the **constellations** layer. Click the **boundaries** check box to outline constellation boundaries. Now zoom in until the constellation almost fills your screen.

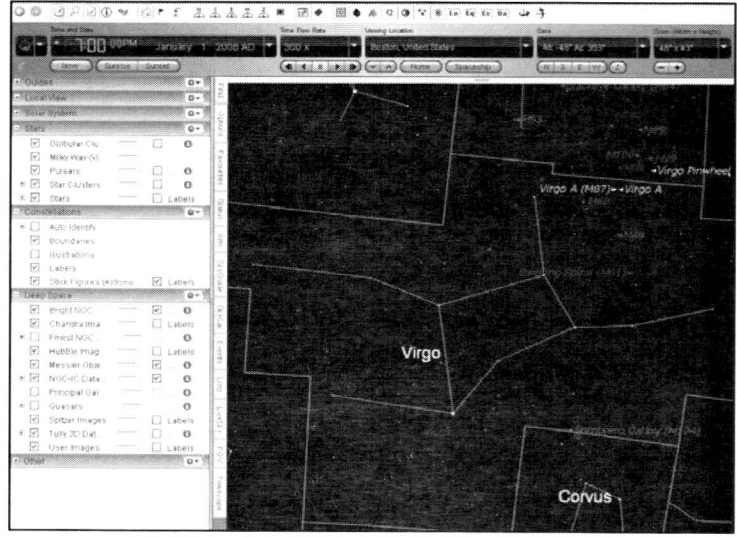

[OPTIONAL] You can increase the number of objects being labeled by Starry Night by expanding the **deep space** and **other** layers of the OPTIONS **side pane**, hovering your pointer over the words **bright NGC objects**, and then clicking to change viewing options. Slide the **labels** bar all the way to the right, then click **OK**. Do the same with **Messier objects**. Follow the same procedure for the **NGC-IC** database, however, slide the **number of objects** bar to the right (don't worry about the **labels** bar in this case since it will be turned off). This will increase the number of objects shown and labeled on your screen. Although the **bright NGC objects** and **Messier objects** should be labeled (right check box), be sure not to select the labeling option for the **NGC-IC database**.

PART 3: VIEWING OBJECTS OF INTEREST IN VIRGO

Here are a few interesting objects found in this constellation. There are, of course, countless items of interest to look at, but only a representative sample of these has been selected.

M104 (Sombrero Galaxy):

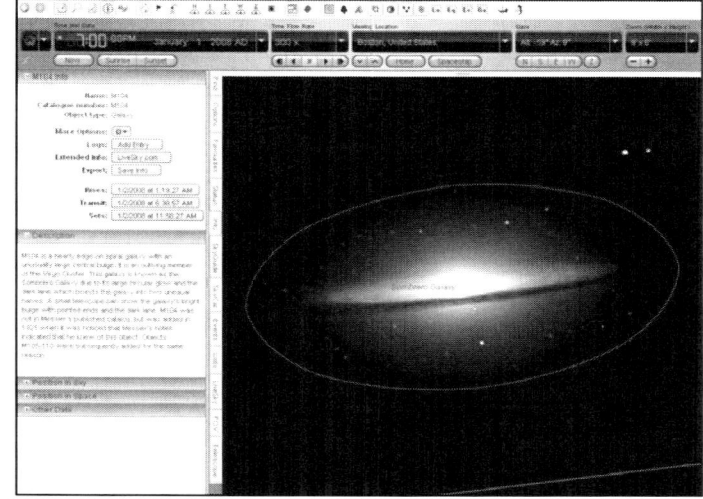

M104 is a galaxy. Galaxies are large collections of stars numbering in the millions, billions, or trillions, and revolving around a common galactic center that most likely includes a supermassive black hole. Galaxies can have many shapes, including elliptical and spiral (such as our own). Galaxies are often millions of light-years apart from each other, with essentially nothing in between.

Look for M104 labeled within the boundaries of the Virgo constellation as shown on your screen.* If an object is not labeled on your screen, hovering your mouse pointer over the object will give you some identifying information. Once you locate what you are looking for, zoom in on the object using the **zoom controls** at the right of the **control panel**. Right click (control click for Mac) on the object and select **show info**, click the plus sign (gray arrow for Mac) to expand all layers, then complete the following table of information.

Date, time:	
Azimuth, altitude:	
Distance from observer (give units):	
Apparent magnitude:	
Angular size (give units—degrees, arc minutes, etc.):	

What do you find most interesting? Indicate some additional info about this object:

*If you have trouble finding M104, click the FIND tab on the left **side pane**. In the search field at the top, type in M104. More than one listing may appear. Click the information icon (**i**) to the right for all entries to learn more about these objects. Grab the right-side double line and click and drag to the right to expand the text box as far as it will go. You will note that the listings are from different databases. This simply means that Starry Night has this object appearing in multiple (different) databases (or lists). All refer to the same object; however, the information given may be slightly different, and the coordinates may vary slightly as well. Double-click on the listing associated with the **Messier objects** database if the object name begins with the letter M. Double-click on the listing associated with the **bright NGC objects** database if the object name begins with the letters NGC. Now zoom in as described previously.

M49 (Galaxy):

M49 is a bright elliptical galaxy located near the center of the Virgo cluster. We will take a closer look at the Virgo cluster later in this activity, and in a separate activity.

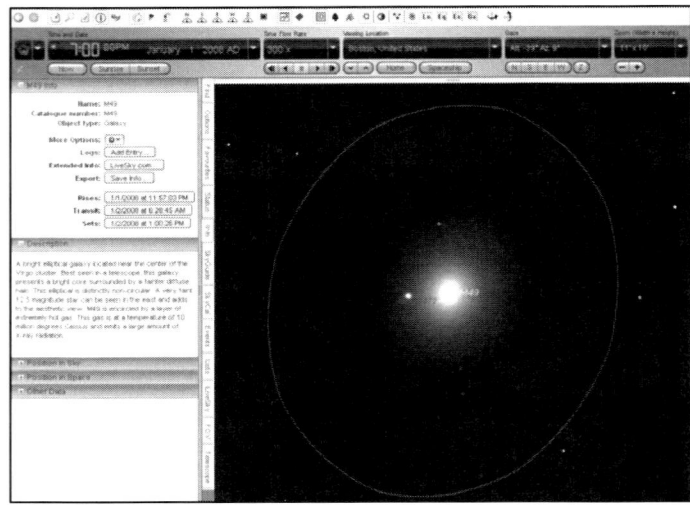

Look for M49 labeled within the boundaries of the Virgo constellation as shown on your screen. If an object is not labeled on your screen, hovering your mouse pointer over the object will give you some identifying information. Once you locate what you are looking for, zoom in on the object using the **zoom controls** at the right of the **control panel**. Right click (control click for Mac) on the object and select INFO, click the plus sign (gray arrow for Mac) to expand all layers, then complete the following table of information. If you have trouble finding M49, click the FIND tab on the left **side pane** and follow the optional procedure given previously.

Date, time:	
Azimuth, altitude:	
Distance from observer (give units):	
Apparent magnitude:	
Angular size (give units—degrees, arc minutes, etc.):	

What do you find most interesting? Indicate some additional info about this object:

VIRGO the MAIDEN (SPRING SKY)

M84 and other Galaxies of the Virgo Cluster:

M84, M86, M87, M58, M89, M90, M60, and M59 all lie in the direction of the Virgo cluster. Many other galaxies of the Virgo cluster can be found next to these on the border between the constellations Virgo and Leo.

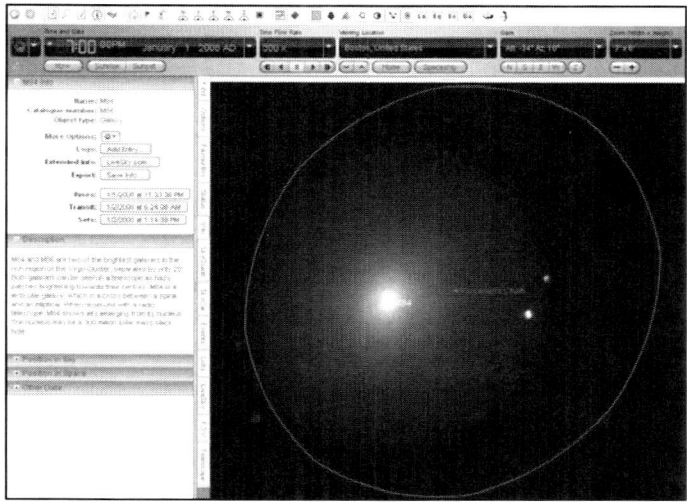

Look for M84 labeled within the boundaries of the Virgo constellation as shown on your screen. If an object is not labeled on your screen, hovering your mouse pointer over the object will give you some identifying information. Once you locate what you are looking for, zoom in on the object using the **zoom controls** at the right of the **control panel**. Right click (control click for Mac) on the object and select **info**, click the plus sign (gray arrow for Mac) to expand all layers, then complete the following table of information. If you have trouble finding M84, click the FIND tab on the left **side pane** and follow the optional procedure given previously. Select two other nearby galaxies and view these as well.

Galaxy:	**M84**		
Date, time:			
Azimuth, altitude:			
Name and catalog number:			
Distance from observer (give units):			
Apparent magnitude:			
Angular size (give units—degrees, arc minutes, etc.):			

What do you find most interesting? Indicate some additional info about these objects:

173

Activity 31—CYGNUS the SWAN (SUMMER SKY)

*These activities are designed to work with the Starry Night software that comes with your text, from any home location you choose, and with the current date and time, unless indicated otherwise. You may always revert to factory default settings by clicking **FILE/ preferences**, then selecting **factory defaults** as needed. You may also undo a command or series of commands on the PC by clicking the **back** button at the top left of the **button bar**. You should refer to the key given at the beginning of this booklet for clarification of "on screen" buttons, controls, and functions. PC **button bar** items can all be accessed through the **menu**. "Right click" on the PC is equivalent to "control click" on the Mac. All activities assume that OpenGL graphics capabilities are enabled on your computer.*

PART 1: THE CYGNUS CONSTELLATION (SUMMER SKY)

In this activity, we are going to explore the constellation Cygnus for objects of interest both for users of the Starry Night software and for amateur astronomers using telescopes in the great outdoors. If you don't have a telescope, this is your chance to explore the night sky as if you did. If you do have a telescope, you can survey possible viewing targets for your nighttime observing sessions. Cygnus is an evening constellation in the summer sky.

There are many different stories from many different cultures associated with each constellation of the night sky. Western culture is most familiar with those whose origin stem from Greek and Roman mythology. One such story claims that Cygnus is Orpheus, the great hero of Thrace who sang and played his lyre so beautifully that wild animals and even the trees would come to hear him. It is said that Orpheus was transported to the sky as a swan so that he could be near his cherished lyre.

Another story has it that Cygnus was the friend of Phaethon, the mortal son of Helios who tried to drive the sun chariot for one day, with disastrous consequences. He was catapulted out of the chariot by Jupiter himself and fell to Earth. His faithful friend Cygnus tried to collect the charred bones to give Phaethon a proper burial. Like a swan, he dived over and over again into the Eridanus River. Jupiter was deeply moved and so rewarded Cygnus by placing him in the sky as Cygnus the swan. Another story describes the love scene between Jupiter and Leda, the wife of Tyndareus, king of Sparta. Jupiter disguised himself as a beautiful swan and swam up to Leda as she bathed in a pool of water. From their union was born Pollux and Helen.

In this activity, we will take a closer look at the Cygnus constellation and examine some of the interesting celestial objects that lie within it.

PART 2: FINDING CYGNUS IN THE SKY

You should begin this activity at sunset. An easy way to do this is to click the drop-down menu to the right of the **date & time** field on the **control panel**, and select **sunset**. Look toward the west by clicking the **W viewing direction** button located on the **button bar** across the top of your screen, or by simply keying in the letter W (Mac users should refer to the button bar commands given at the beginning of this booklet). The screen will pan toward the west. Select a playing speed of **300×** normal time by clicking the drop-down menu at the right of the **time speed** field. Click the STOP **time mode** button when the Sun has set, the stars have come out, and dusk is almost over. Then click on the **constellations** button to show the constellations.

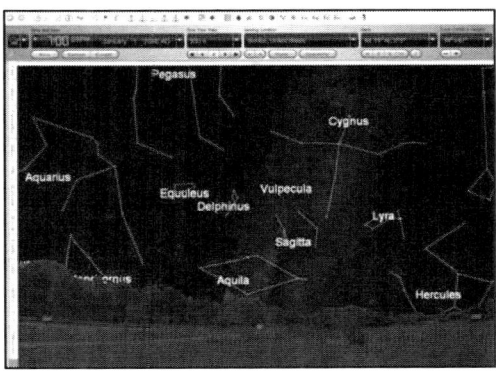

Click the VIEW OPTIONS tab on the left **side pane,** and click the plus sign (gray arrow for Mac) to expand the **deep space** layer. Activate the labels for both the **bright NGC objects** and **Messier objects** by clicking the **labels** box to the right. Then click the information icon **(i)** to the far right to display descriptive information on these Starry Night databases.

You should now see numerous deep-space objects identified on your screen. These are some of the most famous telescope viewing objects in the sky.

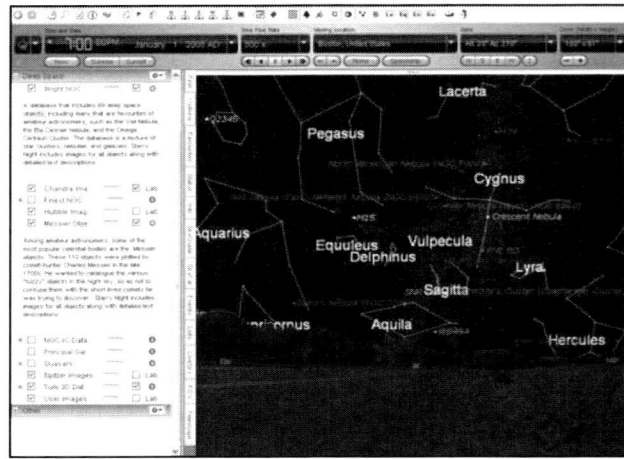

To find the Cygnus constellation, click the FIND tab on the left **side pane**. Type in the name of the constellation you are trying to find. (Hint: If you type the name of the constellation in slowly a list of options will be shown to you.) Don't type out the entire name of the constellation, or you will get an expanded list of all the stars in the constellation. Three or four letters should suffice. From the list shown, double-click the item listed as a CONSTELLATION in the second column (the column that indicates the kind of object being shown). This will center the constellation on your screen and fully illustrate it for you. If the constellation is below the horizon, you may need to select a new time using the **date & time** fields on the **control panel**, and/or turn off the Earth's horizon by clicking the **horizon** button on the **button bar**. If the constellation is up only in the daytime, then you will also need to turn off the Sun by clicking the **daylight** button on the **button bar**. To learn more about this constellation, click the information icon **(i)** to the right of its listing in the FIND tab of the **side pane**.

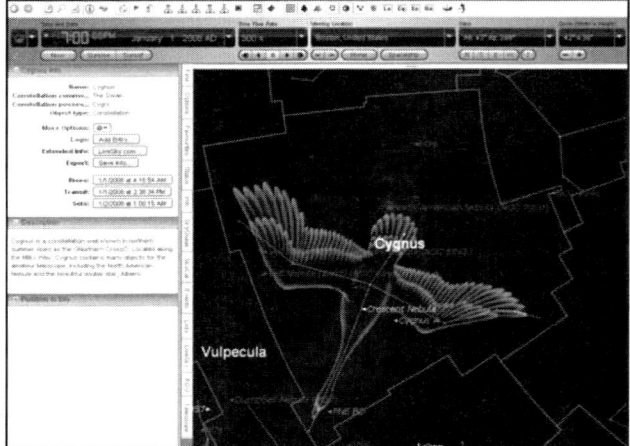

Right click (control click for Mac) on the constellation itself as shown on your screen (careful to not hover over a star or object while doing this) and **deselect** to turn off the full illustration. Click the **galaxies** button on the **button bar** (this turns on the **NGC-IC** database). You should now see only the stick-figure representation of the constellation with little circles and colored dots to indicate deep-space objects of observational interest.

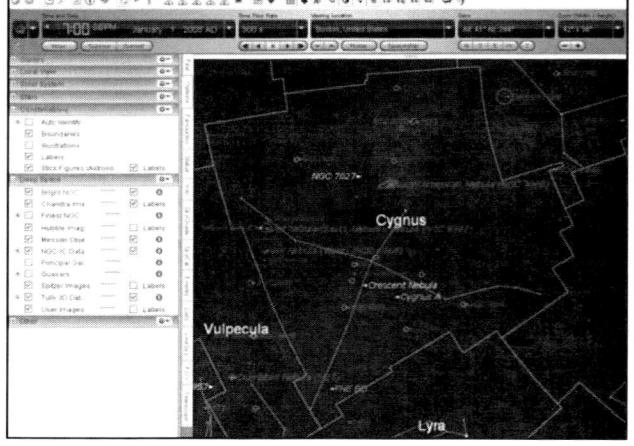

Click the VIEW OPTIONS tab on the left **side pane** and click the plus sign (gray arrow for Mac) to expand the **constellations** layer. Click the **boundaries** check box to outline constellation boundaries. Now zoom in until the constellation almost fills your screen.

[OPTIONAL] You can increase the number of objects being labeled by Starry Night by expanding the **deep space** and **other** layers of the VIEW OPTIONS **side pane**, hovering your pointer over the words **bright NGC objects**, and then clicking to change viewing options. Slide the **labels** bar all the way to the right, then click **OK**. Do the same with **Messier objects**. Follow the same procedure for the **NGC-IC** database, however, slide the **number of objects** bar to the right (don't worry about the **labels** bar in this case since it will be turned off) . This will increase the number of objects shown and labeled on your screen. Although the **bright NGC objects** and **Messier objects** should be labeled (right check box), be sure not to select the labeling option for the **NGC-IC** database.

PART 3: VIEWING OBJECTS OF INTEREST IN CYGNUS

Here are a few interesting objects found in this constellation. There are, of course, countless items of interest to look at, but only a representative sample of these has been selected.

NGC 7000 (North American Nebula):

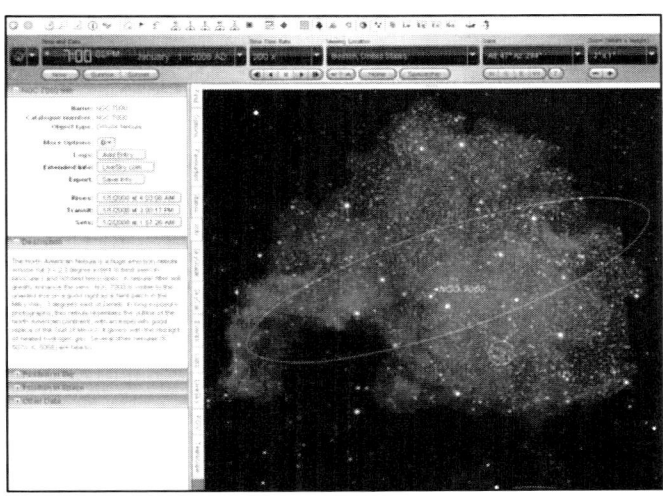

NGC 7000 is an emission/reflection nebula. The term *nebula* initially included all "fuzzy" objects seen in telescopes, but now that many of them have been identified as galaxies, clusters, supernova remnant, planetary nebula, etc. The term *nebula* is mostly used to refer to diffused clouds of interstellar gas and dust that either emit light on their own due to high temperatures (red emission nebula) or reflect light from a nearby star (blue reflection nebula). Often, reflection nebulae accompany emission nebulae. With proper filters and image processing techniques, the color may be an indication of the type of nebular material. The red color is due to hydrogen emission, the blue color is due to reflection of star light from dust particles. Dark nebulae are produced when the dust is so thick that it blocks light from the background stars. It's paradoxical to note that these dark areas are often intense star-forming regions.

Look for NGC 7000 labeled within the boundaries of the Cygnus constellation as shown on your screen.* If an object is not labeled on your screen, hovering your mouse pointer over the object will give you some identifying information. Once you locate what you are looking for, zoom in on the object using the **zoom controls** at the right of the **control panel**. Right click (control click for Mac) on the object and select **show info**, click the plus sign (gray arrow for Mac) to expand all layers, then complete the following table of information.

Date, time:	
Azimuth, altitude:	
Distance from observer (give units):	
Apparent magnitude:	
Angular size (give units—degrees, arc minutes, etc.):	

What do you find most interesting? Indicate some additional info about this object:

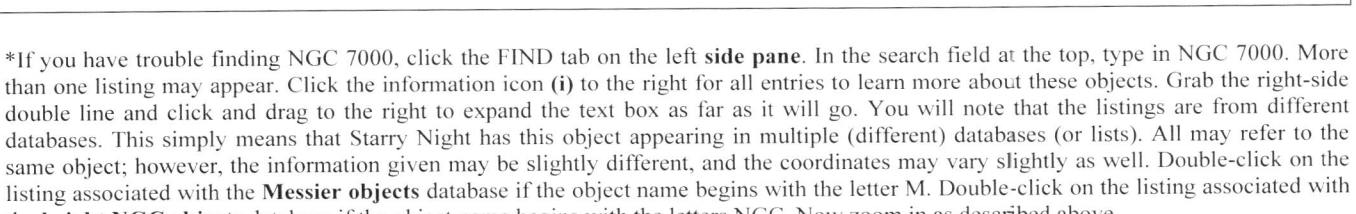

*If you have trouble finding NGC 7000, click the FIND tab on the left **side pane**. In the search field at the top, type in NGC 7000. More than one listing may appear. Click the information icon **(i)** to the right for all entries to learn more about these objects. Grab the right-side double line and click and drag to the right to expand the text box as far as it will go. You will note that the listings are from different databases. This simply means that Starry Night has this object appearing in multiple (different) databases (or lists). All may refer to the same object; however, the information given may be slightly different, and the coordinates may vary slightly as well. Double-click on the listing associated with the **Messier objects** database if the object name begins with the letter M. Double-click on the listing associated with the **bright NGC objects** database if the object name begins with the letters NGC. Now zoom in as described above.

NGC 6960 and NGC 6992 (Veil Nebula / Cygnus Loop):

NGC 6960 and NGC 6992 are supernova remnants. In fact, these two nebulae are opposing filaments of a larger structure referred to as the Cygnus Loop, which makes a circle 2.5 degrees across (five full moons in diameter). Both nebulae are too dim to see with the naked eye; however, they can be viewed with binoculars and/or telescopes. A supernova will occur when a massive star runs out of fuel and becomes unstable, resulting in a powerful explosion that can cause the star to temporarily outshine its host galaxy. When these occur in our own galaxy, they may shine as bright as the Moon and be seen with the naked eye in daytime for weeks and at night for years. Eventually, they dissipate into the interstellar medium, sometimes leaving behind a supernova remnant. Although some supernova remnants are visible optically, many can be observed in other regions of the electromagnetic spectrum. Certain types of supernova may leave behind black holes (detected by X-rays) or spinning neutron stars called pulsars. Thus X-ray sources and pulsars may also be indicators of supernovae.

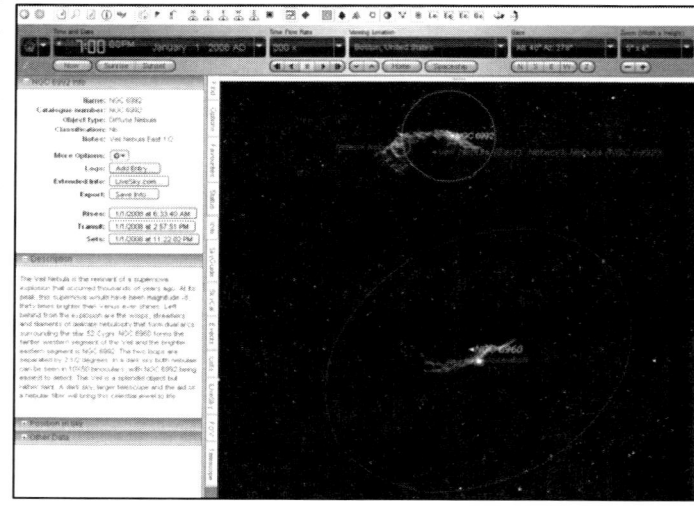

Look for NGC 6960 labeled within the boundaries of the Cygnus constellation as shown on your screen. If an object is not labeled on your screen, hovering your mouse pointer over the object will give you some identifying information. Once you locate what you are looking for, zoom in on the object using the **zoom controls** at the right of the **control panel**. Right click (control click for Mac) on the object and select **show info**, click the plus sign (gray arrow for Mac) to expand all layers, then complete the following table of information. If you have trouble finding NGC 6960, click the FIND tab on the left **side pane** and follow the optional procedure given previously.

Name and catalog number:	**NGC 6960**	**NGC 6992**
Date, time:		
Azimuth, altitude:		
Distance from observer (give units):		
Apparent magnitude:		
Angular size (give units—degrees, arc minutes, etc.):		

What do you find most interesting? Indicate some additional info about these objects:

M39 (Open Cluster):

M39 is an open cluster. Open clusters are groups of stars that have recently formed from the same interstellar cloud of gas and dust. They are relatively young in age and some can be seen with the naked eye and/or with binoculars.

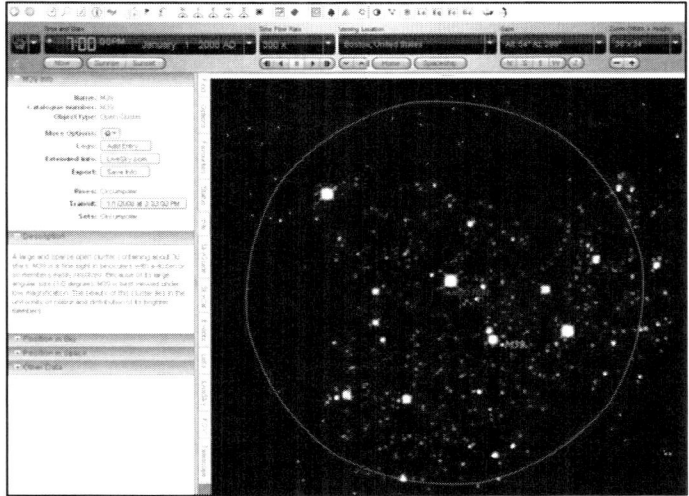

Look for M39 labeled within the boundaries of the Cygnus constellation as shown on your screen. If an object is not labeled on your screen, hovering your mouse pointer over the object will give you some identifying information. Once you locate what you are looking for, zoom in on the object using the **zoom controls** at the right of the **control panel**. Right click (control click for Mac) on the object and select **show info**, click the plus sign (gray arrow for Mac) to expand all layers, then complete the following table of information. If you have trouble finding M39, click the FIND tab on the left **side pane** and follow the optional procedure given previously.

Date, time:	
Azimuth, altitude:	
Distance from observer (give units):	
Apparent magnitude:	
Angular size (give units—degrees, arc minutes, etc.):	

What do you find most interesting? Indicate some additional info about this object:

NGC 7027 (Planetary Nebula):

NGC 7027 is a planetary nebula. Planetary nebulae are formed when a relatively small star, such as our Sun, runs out of fuel. The core of the star collapses to form a very hot and dense white dwarf star. An outer shell of gas is "blown off" during the collapse, producing the spectacular "ring" or "shell" that you see. Try to identify the central white dwarf. Although they are hot and bright, their small size limits our ability to see them. For example, a star the size of our Sun will collapse down to a white dwarf about the size of the Earth, a hundredth the diameter and only one-millionth the volume. Although the size of the star is greatly reduced, the mass stays about the same (except for the gasses ejected into the shell), so white dwarfs are incredibly dense objects, over a million times more dense than that of our Sun.

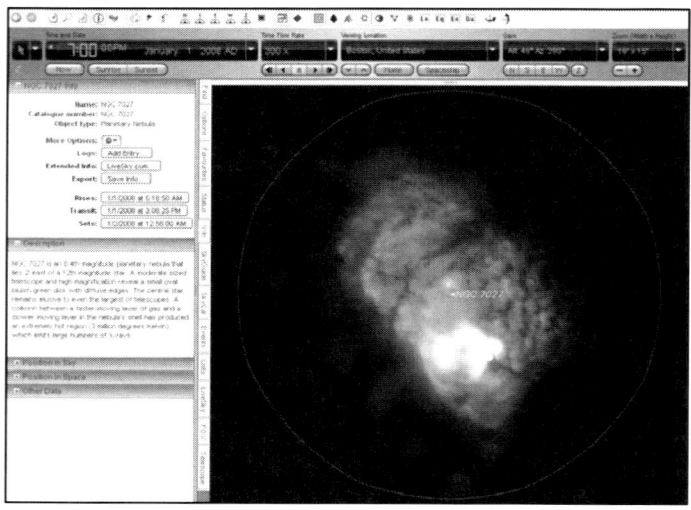

Look for NGC 7027 labeled within the boundaries of the Cygnus constellation as shown on your screen. If an object is not labeled on your screen, hovering your mouse pointer over the object will give you some identifying information. Once you locate what you are looking for, zoom in on the object using the **zoom controls** at the right of the **control panel**. Right click (control click for Mac) on the object and select **show info**, click the plus sign (gray arrow for Mac) to expand all layers, then complete the following table of information. If you have trouble finding NGC 7027, click the FIND tab on the left **side pane** and follow the optional procedure given previously.

Date, time:	
Azimuth, altitude:	
Distance from observer (give units):	
Apparent magnitude:	
Angular size (give units—degrees, arc minutes, etc.):	

What do you find most interesting? Indicate some additional info about this object:

NGC 6826 (Blinking Planetary Nebula):

NGC 6826 is a planetary nebula. The "blinking" effect is caused by the bright central star (white dwarf) overpowering the eye, causing you to not see the surrounding nebulosity. When you look away, your peripheral vision, which is more sensitive to light, sees the nebular shell. In this way, the nebula appears to "blink" when viewed optically through a telescope.

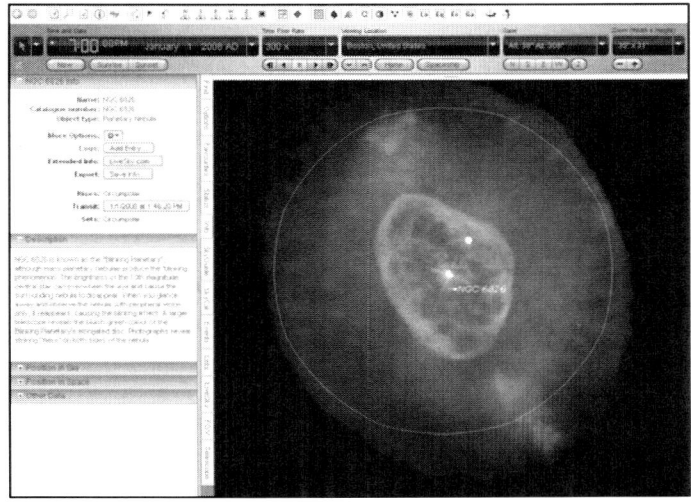

Look for NGC 6826 labeled within the boundaries of the Cygnus constellation as shown on your screen. If an object is not labeled on your screen, hovering your mouse pointer over the object will give you some identifying information. Once you locate what you are looking for, zoom in on the object using the **zoom controls** at the right of the **control panel**. Right click (control click for Mac) on the object and select **show info**, click the plus sign (gray arrow for Mac) to expand all layers, then complete the following table of information. If you have trouble finding NGC 6826, click the FIND tab on the left **side pane** and follow the optional procedure given previously.

Date, time:	
Azimuth, altitude:	
Distance from observer (give units):	
Apparent magnitude:	
Angular size (give units—degrees, arc minutes, etc.):	

What do you find most interesting? Indicate some additional info about this object:

NGC 6946 (Galaxy):

NGC 6946 is a galaxy. Galaxies are large collections of stars numbering in the millions, billions, or trillions, and often revolving around a common galactic center, which most likely includes a supermassive black hole. Galaxies can have many shapes, including spherical and spiral (such as our own). Galaxies are often millions of light-years apart from each other, with essentially nothing in between.

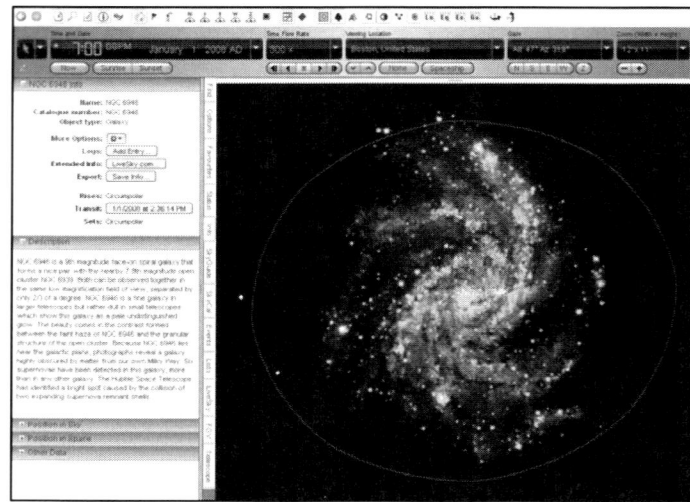

Look for NGC 6946 labeled within the boundaries of the Cygnus constellation as shown on your screen. If an object is not labeled on your screen, hovering your mouse pointer over the object will give you some identifying information. Once you locate what you are looking for, zoom in on the object using the **zoom controls** at the right of the **control panel**. Right click (control click for Mac) on the object and select INFO, click the plus sign (gray arrow for Mac) to expand all layers, then complete the following table of information. If you have trouble finding NGC 6946, click the FIND tab on the left **side pane** and follow the optional procedure given previously.

Date, time:	
Azimuth, altitude:	
Distance from observer (give units):	
Apparent magnitude:	
Angular size (give units—degrees, arc minutes, etc.):	

What do you find most interesting? Indicate some additional info about this object:

Activity 32—ANDROMEDA the PRINCESS (FALL SKY)

*These activities are designed to work with the Starry Night software that comes with your text, from any home location you choose, and with the current date and time, unless indicated otherwise. You may always revert to factory default settings by clicking **FILE/ preferences**, then selecting **factory defaults** as needed. You may also undo a command or series of commands on the PC by clicking the **back** button at the top left of the **button bar**. You should refer to the key given at the beginning of this booklet for clarification of "on screen" buttons, controls, and functions. PC **button bar** items can all be accessed through the **menu**. "Right click" on the PC is equivalent to "control click" on the Mac. All activities assume that OpenGL graphics capabilities are enabled on your computer.*

PART 1: THE ANDROMEDA CONSTELLATION (FALL SKY)

In this activity, we are going to explore the constellation Andromeda for objects of interest both for users of the Starry Night software and for amateur astronomers using telescopes in the great outdoors. If you don't have a telescope, this is your chance to explore the night sky as if you did. If you do have a telescope, you can survey possible viewing targets for your nighttime observing sessions. Andromeda is an evening constellation in the fall sky.

There are many different stories from many different cultures associated with each constellation of the night sky. Western culture is most familiar with those whose origin stem from Greek and Roman mythology. Andromeda is named after the daughter of Cassiopeia and Cepheus, rulers of the ancient land of Ethiopia. When Cassiopeia boasted that she was more beautiful than the Nereids, the water nymphs, the angry sea god Poseidon sent the monster Cetus to ravage the kingdom. Advised by an oracle that the sacrifice of their daughter Andromeda to Cetus was the only way to appease Poseidon, the king and queen duly chained Andromeda to a rock by the sea. Perseus came to the rescue, though, swooping down on the winged horse Pegasus. Perseus had just battled the Medusa, and was able to save Andromeda by showing Medusa's head to Cetus, who instantly turned to stone. In another version, Andromeda was to be sacrificed due to her mother's boasting of her own beauty, and Perseus, flying with the aid of Mercury's winged sandals, swooped down for a rescue. In this version, Perseus had her parents promise him her hand in marriage as Cetus the monster approached, then at the last minute, he struck down the monster by his own hand and with his own sword. The story of Andromeda is thought to be very old and dates back to ancient Mesopotamia. The Chinese placed a sandal in this part of the sky, a reminder that this was the season for making footwear. Another Chinese asterism describes these stars as the Southern Camp Gate, the tongues of two chariots drawn together to make a gate.

In this activity, we will take a closer look at the Andromeda constellation and examine some of the interesting celestial objects that lie within it.

PART 2: FINDING ANDROMEDA IN THE SKY

You should begin this activity at sunset. An easy way to do this is to click the drop-down menu to the right of the **date & time** field on the **control panel**, and select **sunset**. Look toward the west by clicking the **W viewing direction** button located on the **button bar** across the top of your screen, or by simply keying in the letter W (Mac users should refer to the button bar commands given at the beginning of this booklet). The screen will pan toward the west. Select a playing speed of **300×** normal time by clicking the drop-down menu at the right of the **time speed** field. Click the STOP **time mode** button when the Sun has set, the stars have come out, and dusk is almost over. Then click on the **constellations** button to show the constellations.

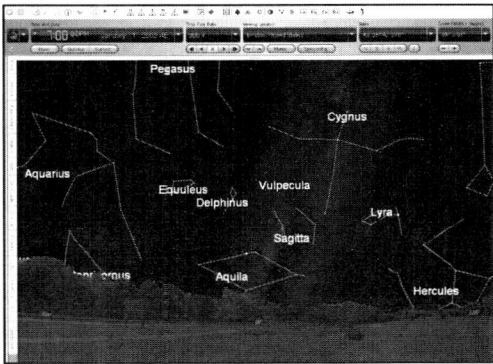

Click the OPTIONS tab on the left **side pane** and expand the **deep space** layer by clicking on the small plus sign to the left (gray arrow for Mac). Activate the labels for both the **bright NGC objects** and **Messier objects** by clicking the **labels** box to the right. Then click the information icon **(i)** to the far right to display descriptive information on these Starry Night databases.

You should now see numerous deep space objects identified on your screen.

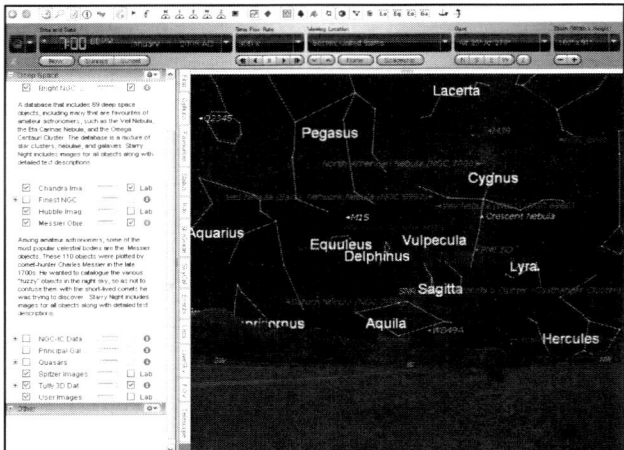

To find the Andromeda constellation, click the FIND tab on the left **side pane** at the left of your screen. Type in the name of the constellation you are trying to find. Type the name in slowly and a list of options will be shown. Don't type out the entire name of the constellation, or you will get an expanded list of all the stars in the constellation. Three or four letters should suffice. From the list shown, double-click the item listed as a CONSTELLATION in the second column (the column that indicates the kind of object being shown). This will center the constellation on your screen and fully illustrate it for you. If the constellation is below the horizon, you may need to select a new time using the **date & time** fields on the **control panel**, and/or turn off the Earth's horizon by clicking the **horizon** button on the **button bar**. If the constellation is up only in the daytime, then you will also need to turn off the Sun by clicking the **daylight** button on the **button bar**. To learn more about this constellation, click the information icon **(i)** to the right of its listing in the FIND tab of the **side pane**.

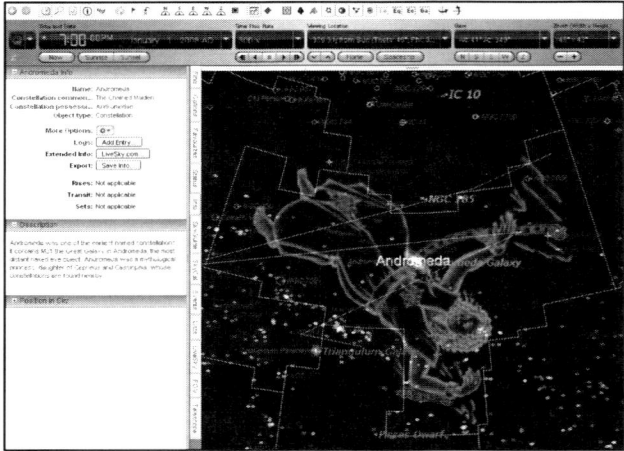

Right click (control click for Mac) on the constellation itself as shown on your screen (careful not to hover over a star or object while doing this) and **deselect** to turn off the full illustration. Click the **galaxies** button on the **button bar** (this turns on the NGC-IC database). You should now see only the stick-figure representation of the constellation with little circles and colored dots to indicate deep-space objects of observational interest.

Click the VIEW OPTIONS tab on the left **side pane** and click the plus sign (gray arrow for Mac) to expand the **constellations** layer. Click the **boundaries** check box to outline constellation boundaries. Now zoom in until the constellation almost fills your screen.

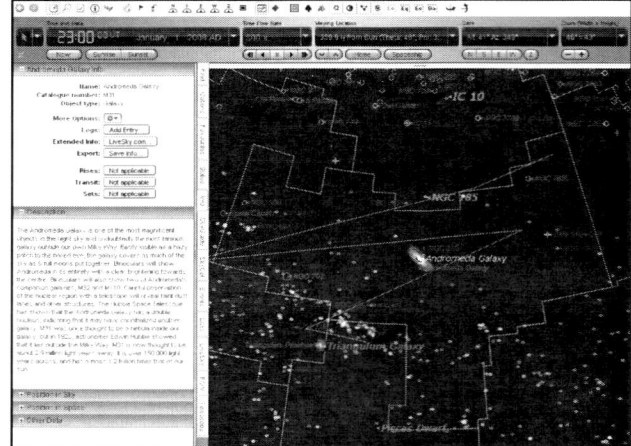

[OPTIONAL] You can increase the number of objects being labeled by Starry Night by expanding the **deep space** and **other** layers of the VIEW OPTIONS **side pane**, hovering your pointer over the words **bright NGC objects**, and then clicking to change viewing options. Slide the **labels** bar all the way to the right, then click **OK**. Do the same with **Messier objects**. Follow the same procedure for the **NGC-IC database**, however, slide the **show** bar to the right (don't worry about the **labels** bar in this case since it will be turned off). This will increase the number of objects shown and labeled on your screen. Although the **bright NGC objects** and **Messier objects** should be labeled (right check box), be sure not to select the labeling option for the **NGC-IC database**.

PART 3: VIEWING OBJECTS OF INTEREST IN ANDROMEDA

Here are a few interesting objects found in this constellation. There are, of course, countless items of interest to look at, but only a representative sample of these has been selected.

NGC 7662 (Blue Snowball Planetary Nebula):

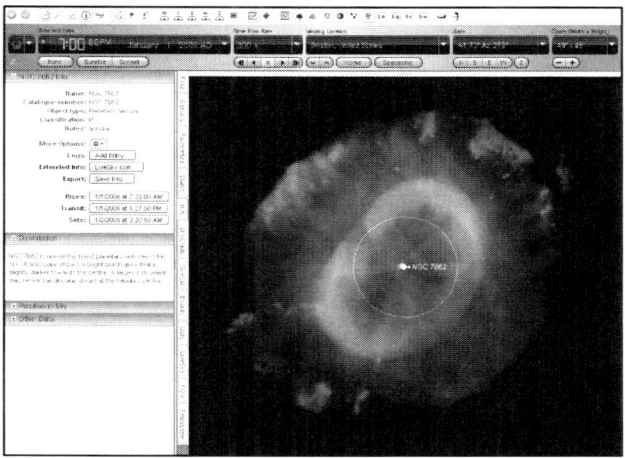

NGC 7662 is a planetary nebula. Planetary nebulae are formed when a relatively small star, such as our Sun, runs out of fuel. The core of the star collapses to form a very hot and dense white dwarf star. An outer shell of gas is "blown off" during the collapse, producing the spectacular "ring" or "shell" that you see. Try to identify the central white dwarf. Although they are hot and bright, their small size limits our ability to see them. For example, a star the size of our Sun will collapse down to a white dwarf about the size of the Earth, a hundredth the diameter and only one-millionth the volume. Although the size of the star is greatly reduced, the mass stays about the same (except for the gasses ejected into the shell), so white dwarfs are incredibly dense objects, over a million times more dense than that of our Sun.

Look for NGC 7662 labeled within the boundaries of the Andromeda constellation as shown on your screen. If an object is not labeled on your screen, hovering your mouse pointer over the object will give you some identifying information. Once you locate what you are looking for, zoom in on the object using the **zoom controls** at the right of the **control panel**. Right click (control click for Mac) on the object and select **show info**, click the plus sign (gray arrow for Mac) to expand all layers, then complete the following table of information. If you have trouble finding NGC 7662, click the FIND tab on the left **side pane** and follow the optional procedure given previously.

Date, time:	
Azimuth, altitude:	
Distance from observer (give units):	
Apparent magnitude:	
Angular size (give units—degrees, arc minutes, etc.):	

What do you find most interesting? Indicate some additional info about this object:

M31 (Andromeda Galaxy):

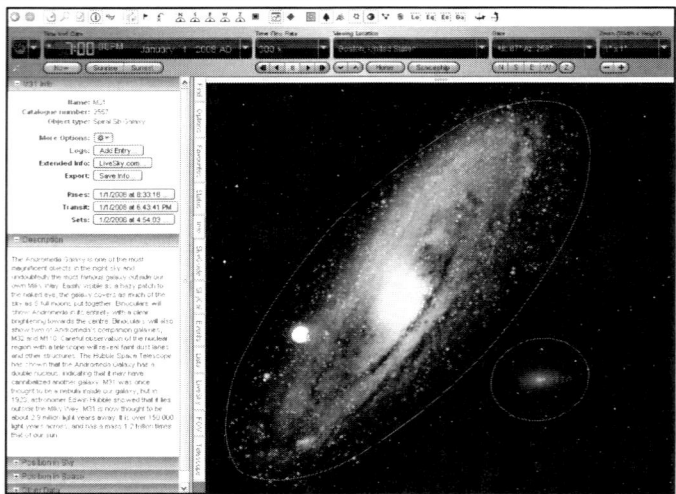

M31 is a galaxy. Galaxies are large collections of stars numbering in the millions, billions, or trillions, and often revolving around a common galactic center that most likely includes a supermassive black hole. Galaxies can have many shapes, including elliptical and spiral (such as our own). Galaxies are often millions of light-years apart from each other, with essentially nothing in between.

Look for M31 labeled within the boundaries of the Andromeda constellation as shown on your screen.* If an object is not labeled on your screen, hovering your mouse pointer over the object will give you some identifying information. Once you locate what you are looking for, zoom in on the object using the **zoom controls** at the right of the **control panel**. Right click (control click for Mac) on the object and select **show info**, click the plus sign (gray arrow for Mac) to expand all layers, then complete the following table of information.

Date, time:	
Azimuth, altitude:	
Distance from observer (give units):	
Apparent magnitude:	
Angular size (give units—degrees, arc minutes, etc.):	

What do you find most interesting? Indicate some additional info about this object:

*If you have trouble finding M31, click the FIND tab on the left **side pane**. In the search field at the top, type in M31. More than one listing may appear. Click the information icon **(i)** to the right for all entries to learn more about these objects. Grab the right-side double line and click and drag to the right to expand the text box as far as it will go. You will note that the listings are from different databases. This simply means that Starry Night has this object appearing in multiple (different) databases (or lists). All refer to the same object; however, the information given may be slightly different, and the coordinates may vary slightly as well. Double-click on the listing associated with the **Messier objects** database if the object name begins with the letter M. Double-click on the listing associated with the **bright NGC objects** database if the object name begins with the letters NGC. Now zoom in as described above.

M32 (Satellite Galaxy of Andromeda):

M32 is a small and compact satellite galaxy of M31. Look for M32 near M31, labeled within the boundaries of the Andromeda constellation as shown on your screen. You may need to zoom in towards M31 for it to show. If an object is not labeled on your screen, hovering your mouse pointer over the object will give you some identifying information.

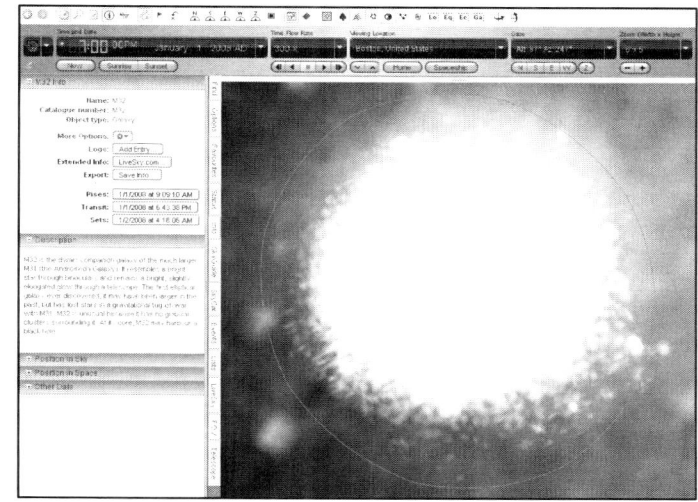

Once you locate what you are looking for, zoom in on the object using the **zoom controls** at the right of the **control panel**. Right click (control click for Mac) on the object and select **info**, click the plus sign (gray arrow for Mac) to expand all layers, then complete the following table of information. If you have trouble finding M32, click the FIND tab on the left **side pane** and follow the optional procedure given previously.

Date, time:	
Azimuth, altitude:	
Distance from observer (give units):	
Apparent magnitude:	
Angular size (give units—degrees, arc minutes, etc.):	

What do you find most interesting? Indicate some additional info about this object:

M110 (Satellite Galaxy of Andromeda):

M110 is a larger and more diffuse satellite galaxy of M31. Look for M110 labeled within the boundaries of the Andromeda constellation as shown on your screen. If an object is not labeled on your screen, hovering your mouse pointer over the object will give you some identifying information. Once you locate what you are looking for, zoom in on the object using the **zoom controls** at the right of the **control panel**. Right click (control click for Mac) on the object and select **info**, click the plus sign (gray arrow for Mac) to expand all layers, then complete the following table of information. If you have trouble finding M110, click the FIND tab on the left **side pane** and follow the optional procedure given previously.

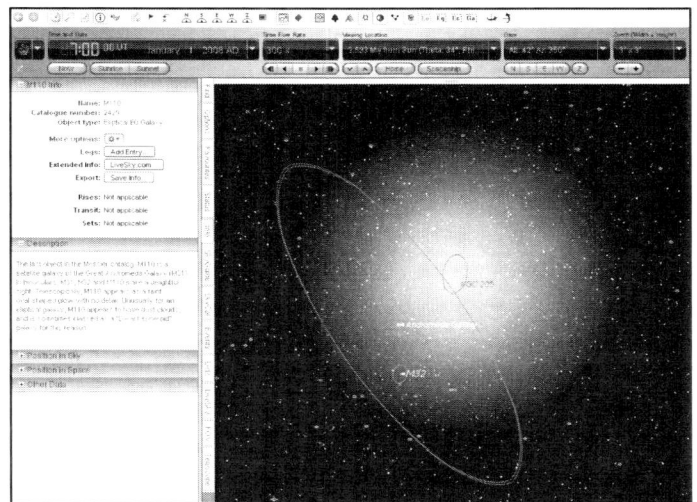

Date, time:	
Azimuth, altitude:	
Distance from observer (give units):	
Apparent magnitude:	
Angular size (give units—degrees, arc minutes, etc.):	

What do you find most interesting? Indicate some additional info about this object:

Activity 33—ORION the MIGHTY HUNTER (WINTER SKY)

*These activities are designed to work with the Starry Night software that comes with your text, from any home location you choose, and with the current date and time, unless indicated otherwise. You may always revert to factory default settings by clicking **FILE/ preferences**, then selecting **factory defaults** as needed. You may also undo a command or series of commands on the PC by clicking the **back** button at the top left of the **button bar**. You should refer to the key given at the beginning of this booklet for clarification of "on screen" buttons, controls, and functions. PC **button bar** items can all be accessed through the **menu**. "Right click" on the PC is equivalent to "control click" on the Mac. All activities assume that OpenGL graphics capabilities are enabled on your computer.*

PART 1: THE ORION CONSTELLATION (WINTER SKY)

In this activity, we are going to explore the constellation Orion for objects of interest both for users of the Starry Night software and for amateur astronomers using telescopes in the great outdoors. If you don't have a telescope, this is your chance to explore the night sky as if you did. If you do have a telescope, you can survey possible viewing targets for your nighttime observing sessions. Orion is an evening constellation in the winter sky.

There are many different stories from many different cultures associated with each constellation of the night sky. Western culture is most familiar with those whose origin stem from Greek and Roman mythology. The Orion constellation is named after Orion the mighty hunter. In one story, Orion was the son of Neptune and the nymph Euryale. In another story, three gods (Zeus, Hermes, and Poseidon) took the hide of a cow, urinated on it, then buried it, from which Orion then rose from ground. Orion was gigantic in size and strength and threatened to hunt and kill all animals on Earth. Gaia, goddess of the Earth, was not pleased with Orion's boast and sent a scorpion to kill him. However, Orion was saved by Ophiuchus, the doctor of antiquity, who gave him an antidote to the scorpion's venom. You can see this scene played out in the stars, for when Scorpius rises in the East, Orion is seen sinking, mortally wounded, to the west. The next night, Orion rises again, restored to health and strength by Ophiuchus; when Scorpius sets to the west, Ophiuchus stands over him, triumphantly.

In another story, Diana, goddess of the hunt, and Aurora, goddess of the dawn, both fell in love with Orion. As Orion sets in the west, his stars fade very slowly as Aurora tries to hold on to him. When he is gone, Aurora weeps bitter tears as can be seen by the dew that collects on the grasses and leaves of the trees. In one version of this story, out of jealousy, Diana shot Orion with an arrow and blinded him. In another version, while Orion was out hunting, Apollo challenged Artemis to shoot at a hare in the bushes. Not realizing that there was no rabbit (it was Orion in the bushes), Artemis shot an arrow and killed him. When she saw what she had done, she was overcome with grief. To honor his spirit, Orion's body was placed in the sky. Even today, her grief can be seen in the sad and cold look that appears on the face of the Moon at night. Other cultures have viewed Orion as an octopus, an animal trap, a Cayman, a turtle, a foot stool, or as the stern of a great canoe.

In this activity, we will take a closer look at the Orion constellation and examine some of the interesting celestial objects that lie within it.

PART 2: FINDING ORION IN THE SKY

You should begin this activity at sunset. An easy way to do this is to click the drop-down menu to the right of the **date & time** field on the **control panel**, and select **sunset**. Look toward the west by clicking the **W viewing direction** button located on the **button bar** across the top of your screen, or by simply keying in the letter W (Mac users should refer to the button bar commands given at the beginning of this booklet). The screen will pan toward the west. Select a playing speed of **300×** normal time by clicking the drop-down menu at the right of the **time speed** field. Click the STOP **time mode** button when the Sun has set, the stars have come out, and dusk is almost over. Then click on the **constellations** button to show the constellations.

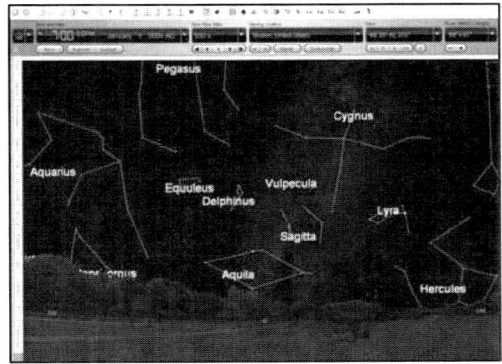

Click the OPTIONS tab on the left **side pane** and expand the **deep space** layer by clicking on the small plus sign to the left (gray arrow for Mac). Activate the labels for both the **bright NGC objects** and **Messier objects** by clicking the **labels** box to the right. Then click the information icon **(i)** to the far right to display descriptive information on these Starry Night databases.

You should now see numerous deep-space objects identified on your screen.

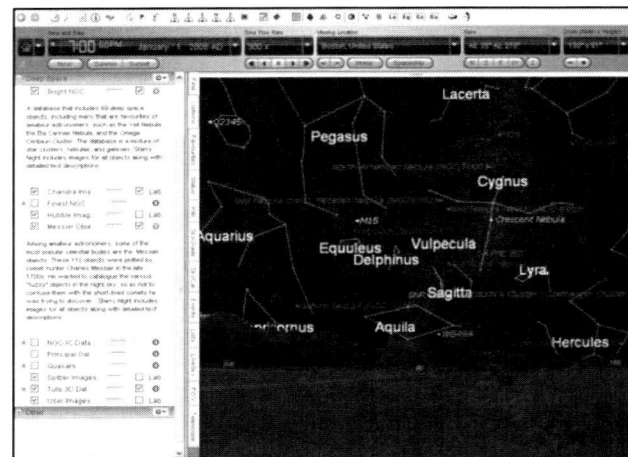

To find the Orion constellation, click the FIND tab on the left **side pane** at the left of your screen. Type in the name of the constellation you are trying to find. (Hint: If you type the name in slowly a list of options will be offered to you.) Don't type out the entire name of the constellation, or you will get an expanded list of all the stars in the constellation. Three or four letters should suffice. From the list shown, double-click the item listed as a **constellation** in the second column (the column that indicates the kind of object being shown). This will center the constellation on your screen and fully illustrate it for you. If the constellation is below the horizon, you may need to select a new time using the **date & time** fields on the **control panel**, and/or turn off the Earth's horizon by clicking the **horizon** button on the **button bar**. If the constellation is up only in the daytime, then you will also need to turn off the Sun by clicking the **daylight** button on the **button bar**. To learn more about this constellation, click the information icon **(i)** to the right of its listing in the FIND tab of the **side pane**.

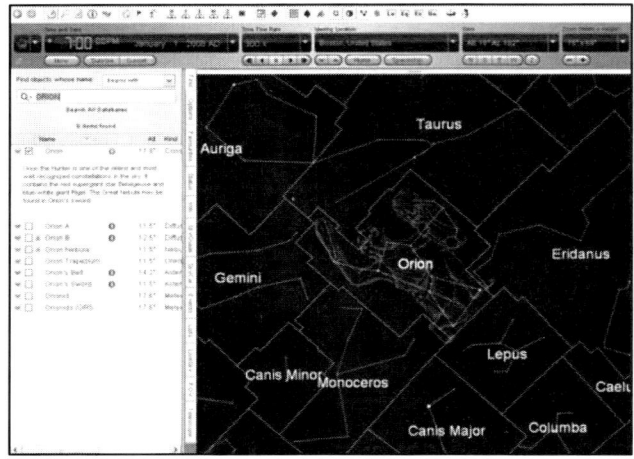

Right click (control click for Mac) on the constellation itself as shown on your screen (being careful to not hover over a star or object while doing this) and **deselect** to turn off the full illustration. Click the **galaxies** button on the **button bar** (this turns on the **NGC-IC** database). You should now see only the stick-figure representation of the constellation with little circles and colored dots to indicate deep-space objects of observational interest.

Click the OPTIONS tab on the left **side pane** and click the plus sign (gray arrow for Mac) to expand the **constellations** layer. Click the **boundaries** check box to outline constellation boundaries. Now zoom in until the constellation almost fills your screen.

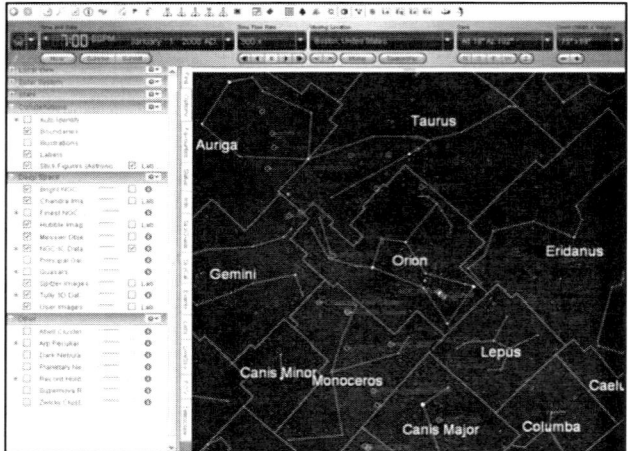

[OPTIONAL] You can increase the number of objects being labeled by Starry Night by expanding the **deep space** and **other** layers of the OPTIONS **side pane**, hovering your pointer over the words **bright NGC objects**, and then clicking to change viewing options. Slide the **labels** bar all the way to the right, then click **OK**. Do the same with **Messier objects**. Follow the same procedure for the **NGC-IC** database, however, slide the **number of objects** bar to the right (don't worry about the **labels** bar in this case since it will be turned off). This will increase the number of objects shown and labeled on your screen. Although the **bright NGC objects** and **Messier objects** should be labeled (right check box), be sure not to select the labeling option for the **NGC-IC** database.

PART 3: VIEWING OBJECTS OF INTEREST IN ORION

Here are a few interesting objects found in this constellation. There are, of course, countless items of interest to look at, but only a representative sample of these has been selected.

<u>M42, M43, and NGC 1977 (Orion Nebula and vicinity):</u>

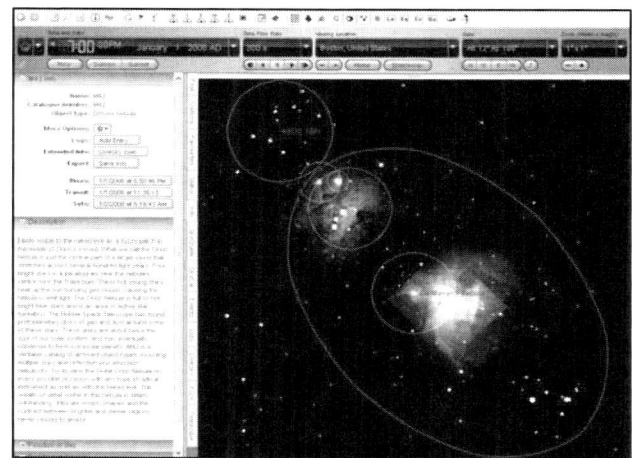

M42, M43, or NGC 1977 are all emission/reflection nebulae. The term *nebula* initially included all "fuzzy" objects seen in telescopes, but now that many of them have been identified as galaxies, clusters, supernovae, planetary nebulae, etc. the term *nebula* is mostly used to refer to diffused clouds of interstellar gas and dust that either emit light on their own due to high temperatures (emission nebulae) or reflect light from a nearby star (reflection nebulae). Often, reflection nebulae accompany emission nebulae, as with these examples here where we see both blue reflection nebulae and red emission nebulae together. The red color is due to hydrogen emission, the blue color is due to reflection of the light of a star from dust particles. Dark nebulae are produced when the dust is so thick that it blocks light from the background stars. It's paradoxical to note that these dark areas are often intense star-forming regions. In fact, the Orion nebula can be thought of as an interstellar nursery, where infant stars are being formed. Star formation regions such as this will eventually evolve into an open cluster.

Look for M42 labeled within the boundaries of the Orion constellation as shown on your screen.* Take note of M43 (de Mairan's nebula) right next to M42. In fact, M43 is part of M42, but separated by a dark band of dust. NGC 1977 is part of a nebulous complex a little farther away. If an object is not labeled on your screen, hovering your mouse pointer over the object will give you some identifying information. Once you locate what you are looking for, zoom in on the object using the **zoom controls** at the right of the **control panel**. Right click (control click for Mac) on the object and select **show info**, click the plus sign (gray arrow for Mac) to expand all layers, then complete the following table of information.

Name and catalog number:	**M42**	**M43**	**NGC 1977**
Date, time:			
Azimuth, altitude:			
Distance from observer (give units):			
Apparent magnitude:			
Angular size (give units—degrees, arc minutes, etc.):			

What do you find most interesting? Indicate some additional info about these objects:

*If you have trouble finding M42, M43, or NGC 1977, click the FIND tab on the left **side pane**. In the search field at the top, type in M42. More than one listing may appear. Click the information icon **(i)** to the right for all entries to learn more about these objects. Grab the right side double line and click and drag to the right to expand the text box as far as it will go. You will note that the listings are from different databases. This simply means that Starry Night has this object appearing in multiple (different) databases (or lists). All refer to the same object, however, the information given may be slightly different and the coordinates may vary slightly. Double-click on the listing associated with the **Messier objects** database if the object name begins with the letter M. Double-click on the listing associated with the **bright NGC objects** database if the object name begins with the letters NGC. Now zoom in as described above.

NGC 2024 (Horsehead Nebula region):

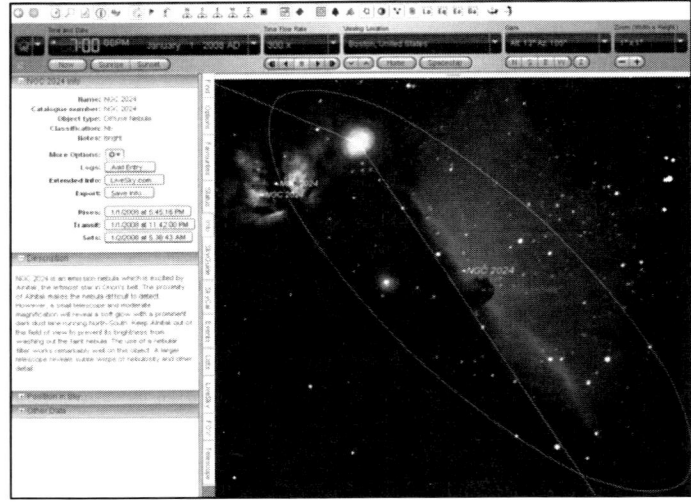

As with the Orion nebula described previously, NGC 2024 and vicinity comprise another nebulous complex of related emission/reflection nebulae. The Horeshead nebula is considered one of the most famous dark nebulae, a region so thick with gas and dust that you can't even see the enshrouded star-forming regions within.

Look for NGC 2024 labeled within the boundaries of the Orion constellation as shown on your screen. If an object is not labeled on your screen, hovering your mouse pointer over the object will give you some identifying information. Once you locate what you are looking for, zoom in on the object using the **zoom controls** at the right of the **control panel**. Right click (control click for Mac) on the object and select **show info**, click the plus sign (gray arrow for Mac) to expand all layers, then complete the following table of information. If you have trouble finding NGC 2024, click the FIND tab on the left **side pane** and follow the optional procedure given previously.

Date, time:	
Azimuth, altitude:	
Distance from observer (give units):	
Apparent magnitude:	
Angular size (give units—degrees, arc minutes, etc.):	

What do you find most interesting? Indicate some additional info about this object:

M78 (Reflection Nebula):

M78 is a reflection nebula that is part of the same cloud of gas and dust that is responsible for M42 (the Orion nebula).

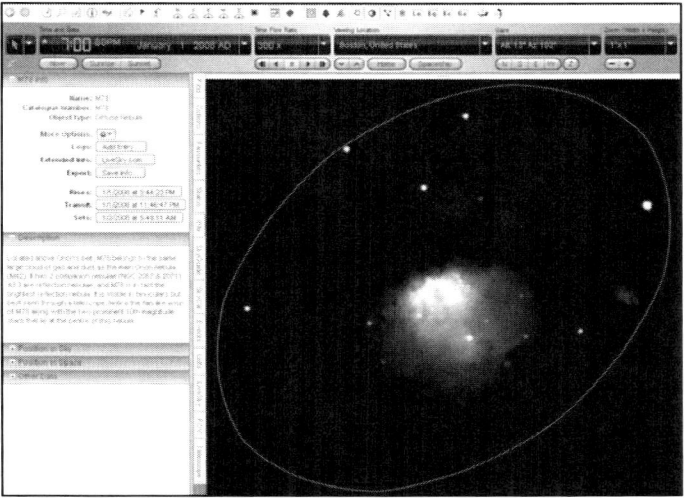

Look for M78 labeled within the boundaries of the Orion constellation as shown on your screen. If an object is not labeled on your screen, hovering your mouse pointer over the object will give you some identifying information. Once you locate what you are looking for, zoom in on the object using the **zoom controls** at the right of the **control panel**. Right click (control click for Mac) on the object and select **show info**, click the plus sign (gray arrow for Mac) to expand all layers, then complete the following table of information. If you have trouble finding M78, click the FIND tab on the left **side pane** and follow the optional procedure given previously.

Date, time:	
Azimuth, altitude:	
Distance from observer (give units):	
Apparent magnitude:	
Angular size (give units—degrees, arc minutes, etc.):	

What do you find most interesting? Indicate some additional info about this object:

Activity 34—URSA MAJOR the GREAT BEAR or BIG DIPPER (NORTHERN HEMISPHERE)

*These activities are designed to work with the Starry Night software that comes with your text, from any home location you choose, and with the current date and time, unless indicated otherwise. You may always revert to factory default settings by clicking **FILE/ preferences**, then selecting **factory defaults** as needed. You may also undo a command or series of commands on the PC by clicking the **back** button at the top left of the **button bar**. You should refer to the key given at the beginning of this booklet for clarification of "on screen" buttons, controls, and functions. PC **button bar** items can all be accessed through the **menu**. "Right click" on the PC is equivalent to "control click" on the Mac. All activities assume that OpenGL graphics capabilities are enabled on your computer.*

PART 1: THE URSA MAJOR CONSTELLATION (NORTHERN SKY)

In this activity, we are going to explore the constellation Ursa Major for objects of interest both for users of the Starry Night software and for amateur astronomers using telescopes in the great outdoors. If you don't have a telescope, this is your chance to explore the night sky as if you did. If you do have a telescope, you can survey possible viewing targets for your nighttime observing sessions. Ursa Major is found in the northern sky near the North Celestial Pole. This means that, depending on your latitude, Ursa Major may be viewable in your evening sky all year round. The farther north you go, the higher Ursa Major will appear in the sky.

There are many different stories from many different cultures associated with each constellation of the night sky. Western culture is most familiar with those whose origin stem from Greek and Roman mythology. In a Greek legend, Zeus was overcome by Callisto's beauty and lay with her. Shortly thereafter she had a son named Arcas. Juno, Zeus's jealous wife, punished Callisto by turning her into a bear. One day, while out hunting, her son, Arcas, not knowing that the bear was his mortal mother, tried to kill her. Zeus rescued Callisto, placing both her and her son in the sky as Ursa Major and Ursa Minor. Juno was terribly upset, so she arranged to keep the bears from ever bathing in the cool, refreshing waters of the sea. This is why both Ursa Major and Ursa Minor circle the North Celestial Pole, and at northern latitudes do not dip below the horizon. Interestingly, other cultures saw a bear in these stars as well. The Native Americans tell of a story where a bear was lost in the forest late at night, a time when it is said that the trees uproot and move around. One of the trees was upset at having been bumped into, and so a great chase commenced, whereupon the bear was thrown into the heavens. Numerous other Native American legends describe warriors, located in the handle of the Big Dipper, each night hunting a different bear, the chase beginning at sunset and ending at sunrise.

Not all stories associated with Ursa Major have to do with bears. Another story tells of how the three stars of the tail represent the three golden apples Hercules had to fetch from the garden of Hesperides. The Babylonians saw a wagon. The Egyptians placed a bull's hind leg in these stars. The Romans saw seven oxen wheeling around the North Star, tilling the fields of heaven. Another version of this indicates that the oxen were driven by Arcturus, assisted by his two dogs, Canes Venatici. Natives of the Mentawai Islands and the Dayak in Borneo saw a hog's jaw. The Chinese saw the god of literature. The Aztecs saw their god Tezcatlipoca, a god of mischief, who had lost a leg (thus, the handle of the big dipper represents only one leg).

In this activity, we will take a closer look at the Ursa Major constellation and examine some of the interesting celestial objects that lie within it.

PART 2: FINDING URSA MAJOR IN THE SKY

You should begin this activity at sunset. An easy way to do this is to click the drop-down menu to the right of the **date & time** field on the **control panel**, and select **sunset**. Look toward the west by clicking the **W viewing direction** button located on the **button bar** across the top of your screen, or by simply keying in the letter W (Mac users should refer to the button bar commands given at the beginning of this booklet). The screen will pan toward the west. Select a playing speed of **300×** normal time by clicking the drop-down menu at the right of the **time speed** field. Click the STOP **time mode** button when the Sun has set, the stars have come out, and dusk is almost over. Then click on the **constellations** button to show the constellations.

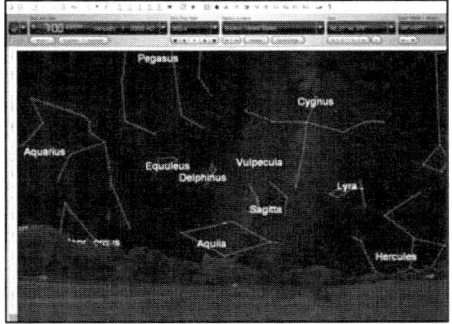

Click the OPTIONS tab on the left **side pane** and expand the **deep space** layer by clicking on the small plus sign to the left (gray arrow for Mac). Activate the labels for both the **bright NGC objects** and **Messier objects** by clicking the **labels** box to the right. Then click the information icon **(i)** to the far right to display descriptive information on these Starry Night databases.

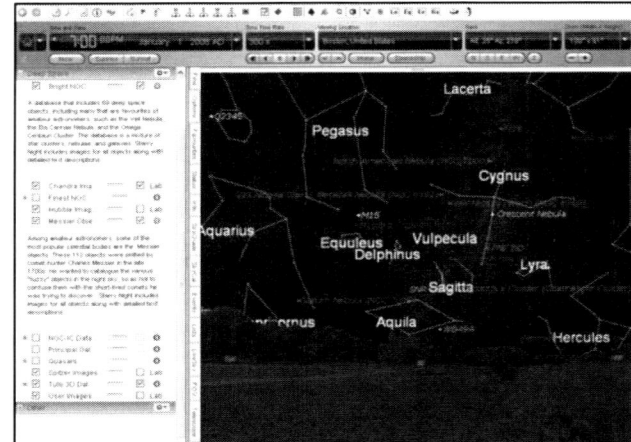

You should now see numerous deep-space objects identified on your screen.

To find the Ursa Major constellation, click the FIND tab on the left **side pane** at the left of your screen. Type in the name of the constellation you are trying to find. (Hint: if you type the name in slowly a list of options will be offered to you.) Don't type out the entire name of the constellation, or you will get an expanded list of all the stars in the constellation. Three or four letters should suffice. From the list shown, double-click the item listed as a CONSTELLATION in the second column (the column that indicates the kind of object being shown). This will center the constellation on your screen and fully illustrate it for you. If the constellation is below the horizon, you may need to select a new time using the **date & time** fields on the **control panel**, and/or turn off the Earth's horizon by clicking the **horizon** button on the **button bar**. If the constellation is up only in the daytime, then you will also need to turn off the Sun by clicking the **daylight** button on the **button bar**. To learn more about this constellation, click the information icon **(i)** to the right of it's listing in the FIND tab of the **side pane**.

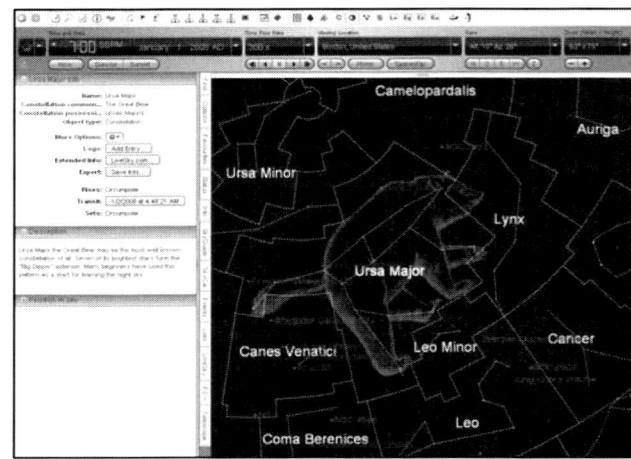

Right click (control click for Mac) on the constellation itself as shown on your screen (careful to not hover over a star or object while doing this) and **deselect** to turn off the full illustration. Click the **galaxies** button on the **button bar** (this turns on the **NGC-IC** database). You should now see only the stick-figure representation of the constellation with little circles and colored dots to indicate deep-space objects of observational interest.

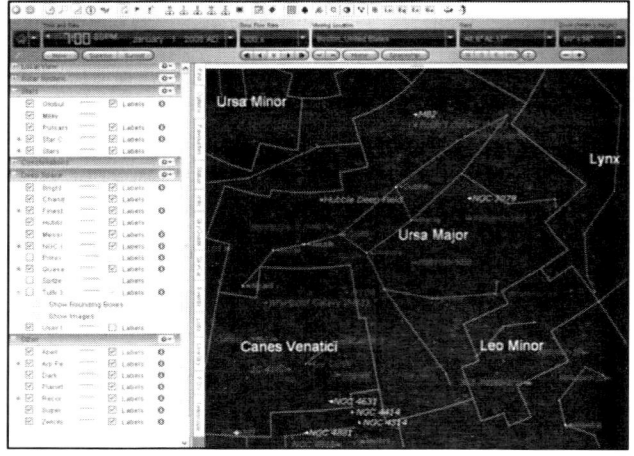

Click the OPTIONS tab on the left **side pane** and click the plus sign (gray arrow for Mac) to expand the **constellations** layer. Click the **boundaries** check box to outline constellation boundaries. Now zoom in until the constellation almost fills your screen.

[OPTIONAL] You can increase the number of objects being labeled by Starry Night by expanding the **deep space** and **other** layers of the OPTIONS **side pane**, hovering your pointer over the words **bright NGC objects**, and then clicking to change viewing options. Slide the **labels** bar all the way to the right, then click **OK**. Do the same with **Messier objects**. Follow the same procedure for the **NGC-IC** database, however, slide the **number of objects** bar to the right (don't worry about the **labels** bar in this case since it will be turned off). This will increase the number of objects shown and labeled on your screen. Although the **bright NGC objects** and **Messier objects** should be labeled (right check box), be sure not to select the labeling option for the **NGC-IC** database.

PART 3: VIEWING OBJECTS OF INTEREST IN URSA MAJOR

Here are a few interesting objects found in this constellation. There are, of course, countless items of interest to look at, but only a representative sample of these has been selected.

<u>M97 (Owl Planetary Nebula):</u>

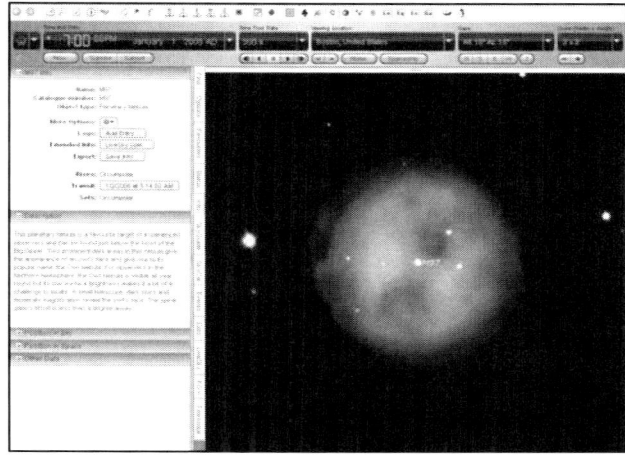

M97 is a planetary nebula. Planetary nebulae are formed when a relatively small star, such as our Sun, runs out of fuel. The core of the star collapses to form a very hot and dense white dwarf star. An outer shell of gas is "blown off" during the collapse, producing the spectacular "ring" or "shell" that you see. Try to identify the central white dwarf. Although they are hot and bright, their small size limits our ability to see them. For example, a star the size of our Sun will collapse down to a white dwarf about the size of the Earth, a hundredth the diameter and only one-millionth the volume. Although the size of the star is greatly reduced, the mass stays about the same (except for the gasses ejected into the shell), so white dwarfs are incredibly dense objects, over a million times more dense than that of our Sun.

Look for M97 labeled within the boundaries of the Ursa Major constellation as shown on your screen.* If an object is not labeled on your screen, hovering your mouse pointer over the object will give you some identifying information. Once you locate what you are looking for, zoom in on the object using the **zoom controls** at the right of the **control panel**. Right click (control click for Mac) on the object and select **show info**, click the plus sign (gray arrow for Mac) to expand all layers, then complete the following table of information.

Date, time:	
Azimuth, altitude:	
Distance from observer (give units):	
Apparent magnitude:	
Angular size (give units—degrees, arc minutes, etc.):	

*If you have trouble finding M97, click the FIND tab on the left **side pane**. In the search field at the top, type in M97. More than one listing may appear. Click the information icon **(i)** to the right for all entries to learn more about these objects. Grab the right-side double line and click and drag to the right to expand the text box as far as it will go. You will note that the listings are from different databases. This simply means that Starry Night has this object appearing in multiple (different) databases (or lists). All may refer to the same object, however, the information given may be slightly different and the coordinates may vary slightly as well. Double-click on the listing associated with the **Messier objects** database if the object name begins with the letter M. Double-click on the listing associated with the **bright NGC objects** database if the object name begins with the letters NGC. Now zoom in as described above.

What do you find most interesting? Indicate some additional info about this object:

M81 (Bode's Galaxy) & M82 (Cigar Galaxy):

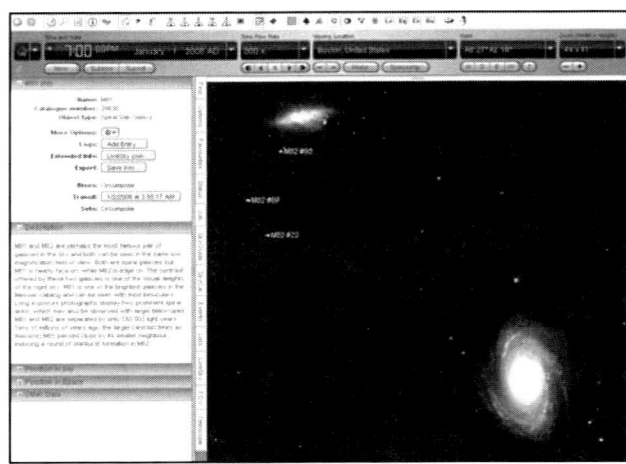

M81 and M82 are galaxies. Galaxies are large collections of stars numbering in the millions, billions, or trillions, revolving around a common galactic center that most likely includes a supermassive black hole. Galaxies can have many shapes, including elliptical and spiral (such as our own). Galaxies are often millions of light-years apart from each other with essentially nothing in between. What gives these two galaxies such different appearances has to do with our viewing angle. M81 is being viewed face-on, while M82 is being viewed from the side. Spiral galaxies will appear elongated when viewed from the side.

Look for M81 labeled within the boundaries of the Ursa Major constellation as shown on your screen. If an object is not labeled on your screen, hovering your mouse pointer over the object will give you some identifying information. Once you locate what you are looking for, zoom in on the object using the **zoom controls** at the right of the **control panel**. Right click (control click for Mac) on the object and select **show info**, click the plus sign (gray arrow for Mac) to expand all layers, then complete the following table of information. If you have trouble finding M81, click the FIND tab on the left **side pane** and follow the optional procedure given previously.

Date, time:		
Azimuth, altitude:		
Distance from observer (give units):		
Apparent magnitude:		
Angular size (give units—degrees, arc minutes, etc.):		

What do you find most interesting? Indicate some additional info about this object:

M108 (Galaxy):

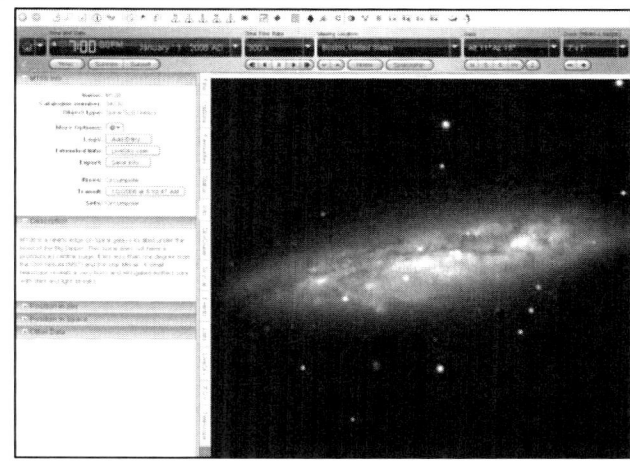

M108 is a galaxy being viewed from the side. This gives it an elongated appearance.

Look for M108 labeled within the boundaries of the Ursa Major constellation as shown on your screen. If an object is not labeled on your screen, hovering your mouse pointer over the object will give you some identifying information. Once you locate what you are looking for, zoom in on the object using the **zoom controls** at the right of the **control panel**. Right click (control click for Mac) on the object and select **show info**, click the plus sign (gray arrow for Mac) to expand all layers, then complete the following table of information. If you have trouble finding M108, click the FIND tab on the left **side pane** and follow the optional procedure given previously.

Date, time:	
Azimuth, altitude:	
Distance from observer (give units):	
Apparent magnitude:	
Angular size (give units—degrees, arc minutes, etc.):	

What do you find most interesting? Indicate some additional info about this object:

M109 (Galaxy):

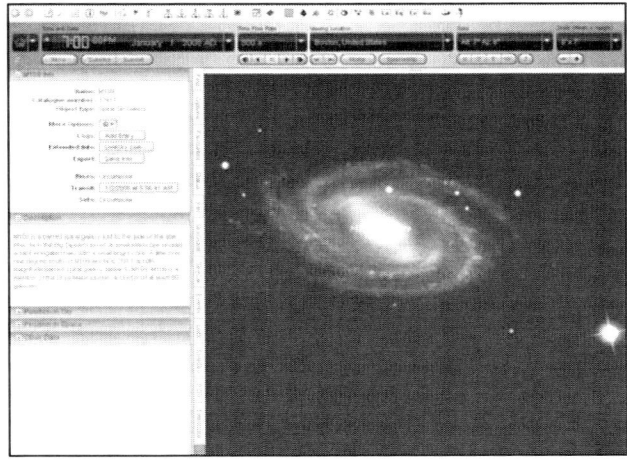

M109 is a barred spiral galaxy viewed almost face-on. It is about 55 million light-years from our galaxy. That means the light takes 55 million years to get to us. We are essentially looking back in time 55 million years when viewing this object.

Look for M109 labeled within the boundaries of the Ursa Major constellation as shown on your screen. If an object is not labeled on your screen, hovering your mouse pointer over the object will give you some identifying information. Once you locate what you are looking for, zoom in on the object using the **zoom controls** at the right of the **control panel**. Right click (control click for Mac) on the object and select **show info**, click the plus sign (gray arrow for Mac) to expand all layers, then complete the following table of information. If you have trouble finding M109, click the FIND tab on the left **side pane** and follow the optional procedure given previously.

Date, time:	
Azimuth, altitude:	
Distance from observer (give units):	
Apparent magnitude:	
Angular size (give units—degrees, arc minutes, etc.):	

What do you find most interesting? Indicate some additional info about this object:

M101 (Pinwheel Galaxy):

M101 is a large spiral face-on galaxy 170,000 light-years across that is half a degree in diameter (same angular size as the full moon). It is 27 million light-years from our galaxy. That means the light takes 27 million years to get to us. We are essentially looking back in time 27 million years when viewing this object.

Look for M101 labeled within the boundaries of the Ursa Major constellation as shown on your screen. If an object is not labeled on your screen, hovering your mouse pointer over the object will give you some identifying information. Once you locate what you are looking for, zoom in on the object using the **zoom controls** at the right of the **control panel**. Right click (control click for Mac) on the object and select **show info**, click the plus sign (gray arrow for Mac) to expand all layers, then complete the following table of information. If you have trouble finding M101, click the FIND tab on the left **side pane** and follow the optional procedure given previously.

Date, time:	
Azimuth, altitude:	
Distance from observer (give units):	
Apparent magnitude:	
Angular size (give units—degrees, arc minutes, etc.):	

What do you find most interesting? Indicate some additional info about this object:

Activity 35—CENTAURUS the CENTAUR (SOUTHERN HEMISPHERE)

*These activities are designed to work with the Starry Night software that comes with your text, from any home location you choose, and with the current date and time, unless indicated otherwise. You may always revert to factory default settings by clicking **FILE/ preferences**, then selecting **factory defaults** as needed. You may also undo a command or series of commands on the PC by clicking the **back** button at the top left of the **button bar**. You should refer to the key given at the beginning of this booklet for clarification of "on screen" buttons, controls, and functions. PC **button bar** items can all be accessed through the **menu**. "Right click" on the PC is equivalent to "control click" on the Mac. All activities assume that OpenGL graphics capabilities are enabled on your computer.*

PART 1: THE CENTAURUS CONSTELLATION
(BEST VIEWED FROM THE SOUTHERN HEMISPHERE)

In this activity, we are going to explore the constellation Centaurus, best viewed from the Southern Hemisphere, for objects of interest both for users of the Starry Night software and for amateur astronomers using telescopes in the great outdoors. If you don't have a telescope, this is your chance to explore the night sky as if you did. If you do have a telescope, you can survey possible viewing targets for your nighttime observing sessions. Centaurus is not easily seen from northern latitudes; however, should you travel south to equatorial regions and/or the Southern Hemisphere, you will find it a prominent constellation of the Southern Hemisphere sky.

There are many different stories from many different cultures associated with each constellation of the night sky. Western culture is most familiar with those whose origin stem from Greek and Roman mythology. These stars most likely represent Chiron, a Centaur, a creature half man and half horse. Most Centaurs were savage and brutal, but Chiron was extremely wise and an educator. Hercules accidentally wounded him, but being immortal, Chiron suffered in great pain but could not die. Zeus mercifully allowed him to die, and placed him among the stars.

In another legend, Ixion, the scorching midday Sun, was in love with Juno, Jupiter's wife. To protect her, Jupiter created a cloud in Juno's image, and when Ixion lay with this cloud, he fathered the Centaurs. In yet another legend, the god Cronos slept with Phylira, a sea nymph. To disguise himself from his wife, Rhea, he took the shape of a horse, and from this union were born the Centaurs.

In this activity, we will take a closer look at the Centaurus constellation and examine some of the interesting celestial objects that lie within it.

PART 2: FINDING CENTAURUS IN THE SKY

You should begin this activity at sunset. An easy way to do this is to click the drop-down menu to the right of the **date & time** field on the **control panel**, and select **sunset**. Look toward the west by clicking the **W viewing direction** button located on the **button bar** across the top of your screen, or by simply keying in the letter W (Mac users should refer to the button bar commands given at the beginning of this booklet). The screen pans toward the west. Select a playing speed of **300×** normal time by clicking the drop-down menu at the right of the **time speed** field. Click the STOP **time mode** button when the Sun has set, the stars have come out, and dusk is almost over. Then click on the **constellations** button to show the constellations.

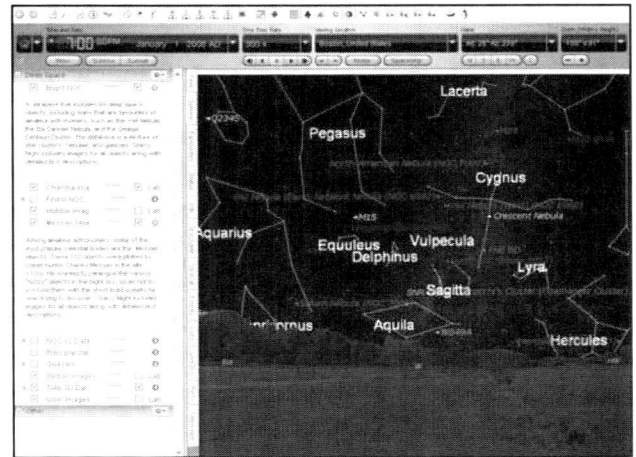

Click the VIEW OPTIONS tab on the left **side pane** and click the plus sign (gray arrow for Mac) to expand the **deep space** layer.

Activate the labels for both the **bright NGC objects** and **Messier objects** by clicking the **labels** box to the right. Then click the information icon **(i)** to the far right to display descriptive information on these Starry Night databases.

You should now see numerous deep-space objects identified on your screen.

To find the Centaurus constellation, click the FIND tab on the left **side pane** at the left of your screen. Type in the name of the constellation you are trying to find. (Hint: If you type the name in slowly a list of options will be offered to you.) Don't type out the entire name of the constellation, or you will get an expanded list of all the stars in the constellation. Three or four letters should suffice. From the list shown, double-click the item listed as a CONSTELLATION in the second column (the column that indicates the kind of object being shown). This will center the constellation on your screen and fully illustrate it for you. If the constellation is below the horizon, you may need to select a new time using the **date & time** fields on the **control panel**, and/or turn off the Earth's horizon by clicking the **horizon** button on the button bar. If the constellation is up only in the daytime, then you will also need to turn off the Sun by clicking the **daylight** button on the **button bar**. To learn more about this constellation, click the information icon **(i)** to the right of its listing in the FIND tab of the **side pane**.

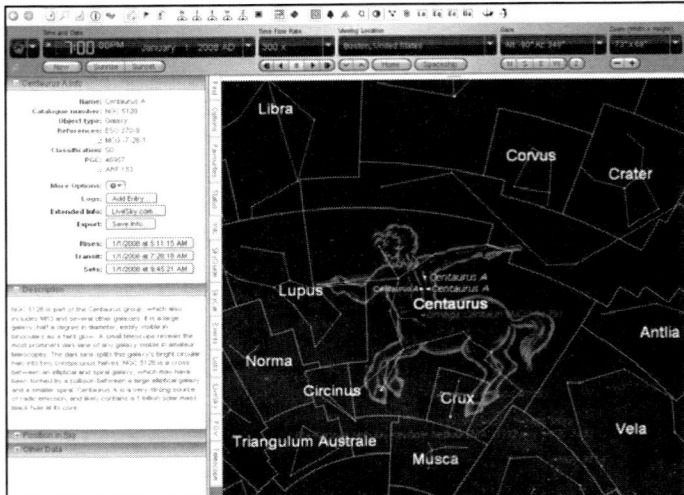

Right click (control click for Mac) on the constellation itself as shown on your screen (careful to not hover over a star or object while doing this) and **deselect** to turn off the full illustration. Click the **galaxies** button on the **button bar** (this turns on the **NGC-IC** database). You should now see only the stick-figure representation of the constellation with little circles and colored dots to indicate deep-space objects of observational interest.

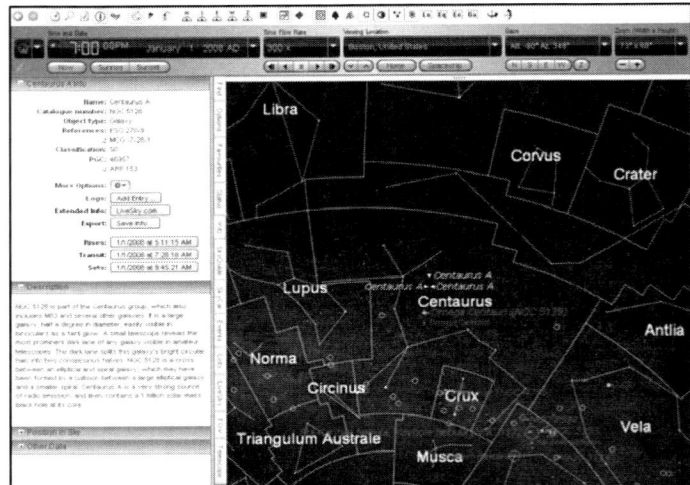

Click the OPTIONS tab on the left **side pane** and click the plus sign (gray arrow for Mac) to expand the **constellations** layer. Click the **boundaries** check box to outline constellation boundaries. Now zoom in until the constellation almost fills your screen.

[OPTIONAL] You can increase the number of objects being labeled by Starry Night by expanding the **deep space** and **other** layers of the OPTIONS **side pane**, hovering your pointer over the words **bright NGC objects**, and then clicking to change viewing options. Slide the **labels** bar all the way to the right, then click **OK**. Do the same with **Messier objects.** Follow the same procedure for the **NGC-IC** database, however, slide the **number of objects** bar to the right (don't worry about the **labels** bar in this case since it will be turned off). This will increase the number of objects shown and labeled on your screen. Although the **bright NGC objects** and **Messier objects** should be labeled (right check box), be sure not to select the labeling option for the **NGC-IC** database.

PART 3: VIEWING OBJECTS OF INTEREST IN CENTAURUS

Here are a few interesting objects found in this constellation. There are, of course, countless items of interest to look at, but only a representative sample of these has been selected.

NGC 5128 (Centaurus A Galaxy):

NGC 5128 is a galaxy. Galaxies are large collections of stars numbering in the millions, billions, or trillions, revolving around a common galactic center that most likely includes a supermassive black hole. Galaxies can have many shapes, including elliptical and spiral (such as our own). Galaxies are often millions of light-years apart from each other, with essentially nothing in between. This galaxy is large, half a degree across (the size of the full moon), and easily visible with binoculars as a faint glow. This galaxy is a cross between an elliptical and a spiral, and may have formed by galactic collision. It is 15 million light-years from our galaxy. That means the light takes 15 million years to get to us. We are essentially looking back in time 15 million years when viewing this object.

Look for NGC 5128 labeled within the boundaries of the Centaurus constellation as shown on your screen.* If an object is not labeled on your screen, hovering your mouse pointer over the object will give you some identifying information. Once you locate what you are looking for, zoom in on the object using the **zoom controls** at the right of the **control panel**. Right click (control click for Mac) on the object and select **show info**, click the plus sign (gray arrow for Mac) to expand all layers, then complete the following table of information.

Date, time:	
Azimuth, altitude:	
Distance from observer (give units):	
Apparent magnitude:	
Angular size (give units—degrees, arc minutes, etc.):	

What do you find most interesting? Indicate some additional info about this object:

*If you have trouble finding NGC 5128, click the FIND tab on the left **side pane**. In the search field at the top, type in NGC 5128. More than one listing may appear. Click the information icon (**i**) to the right for all entries to learn more about these objects. Grab the right-side double line and click and drag to the right to expand the text box as far as it will go. You will note that the listings are from different databases. This simply means that Starry Night has this object appearing in multiple (different) databases (or lists). All may refer to the same object, however, the information given may be slightly different, and the coordinates may vary slightly as well. Double-click on the listing associated with the **Messier objects** database if the object name begins with the letter M. Double-click on the listing associated with the **bright NGC objects** database if the object name begins with the letters NGC. Now zoom in as described above.

NGC 5139 (Omega Centauri Globular Cluster):

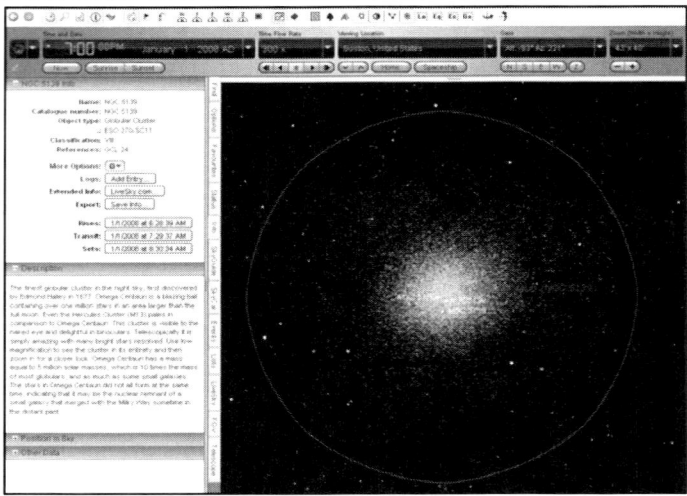

NGC 5139 is a globular cluster. Globular clusters are densely packed groups of stars that look like cotton balls through a telescope. There may be between 50,000 and 2 million stars in each cluster. The stars tend to be very old and the clusters distributed spherically around the center of our galaxy. Astronomers believe they are remnants from an earlier stage of our galaxy's evolution, before our galaxy flattened out to its current shape. This cluster is considered the finest globular cluster in the night sky, consisting of over a million stars in an area the size of the full moon.

Look for NGC 5139 labeled within the boundaries of the Centaurus constellation as shown on your screen. If an object is not labeled on your screen, hovering your mouse pointer over the object will give you some identifying information. Once you locate what you are looking for, zoom in on the object using the **zoom controls** at the right of the **control panel**. Right click (control click for Mac) on the object and select **show info**, click the plus sign (gray arrow for Mac) to expand all layers, then complete the following table of information. If you have trouble finding NGC 5139, click the FIND tab on the left **side pane** and follow the optional procedure given previously.

Date, time:	
Azimuth, altitude:	
Distance from observer (give units):	
Apparent magnitude:	
Angular size (give units—degrees, arc minutes, etc.):	

What do you find most interesting? Indicate some additional info about this object:

NGC 4945 (Active Galaxy):

NGC 4945 is an edge on spiral galaxy. This is a "Seyfert galaxy," a galaxy with an active galactic nuclei, possibly powered by a supermassive black hole.

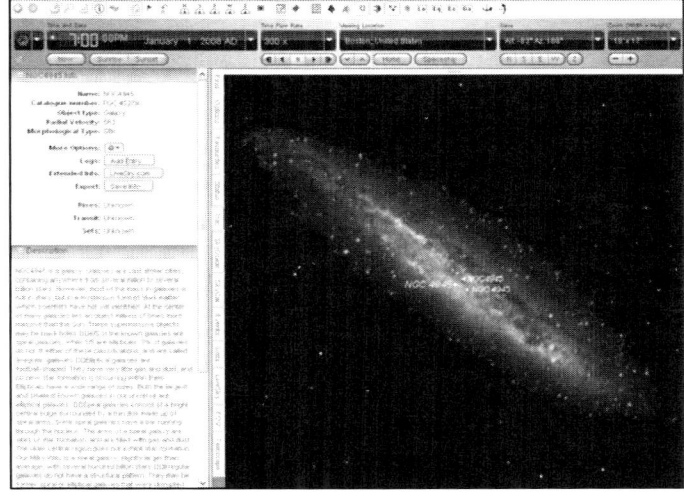

Look for NGC 4945 labeled within the boundaries of the Centaurus constellation as shown on your screen. If an object is not labeled on your screen, hovering your mouse pointer over the object will give you some identifying information. Once you locate what you are looking for, zoom in on the object using the **zoom controls** at the right of the **control panel**. Right click (control click for Mac) on the object and select **show info**, click the plus sign (gray arrow for Mac) to expand all layers, then complete the following table of information. If you have trouble finding NGC 4945, click the FIND tab on the left **side pane** and follow the optional procedure given previously.

Date, time:	
Azimuth, altitude:	
Distance from observer (give units):	
Apparent magnitude:	
Angular size (give units—degrees, arc minutes, etc.):	

Provide a sketch of your own as you see it in Starry Night:

NGC 3918 (Blue Planetary Nebula):

NGC 3918 is a planetary nebula. Planetary nebulae are formed when a relatively small star, such as our Sun, runs out of fuel. The core of the star collapses to form a very hot and dense white dwarf star. An outer shell of gas is "blown off" during the collapse, producing the spectacular "ring" or "shell" that you see. Try to identify the central white dwarf. Although they are hot and bright, their small size limits our ability to see them. For example, a star the size of our Sun will collapse down to a white dwarf about the size of the Earth, a hundredth the diameter and only one-millionth the volume. Although the size of the star is greatly reduced, the mass stays about the same (except for the gasses ejected into the shell), so white dwarfs are incredibly dense objects, over a million times more dense than that of our Sun.

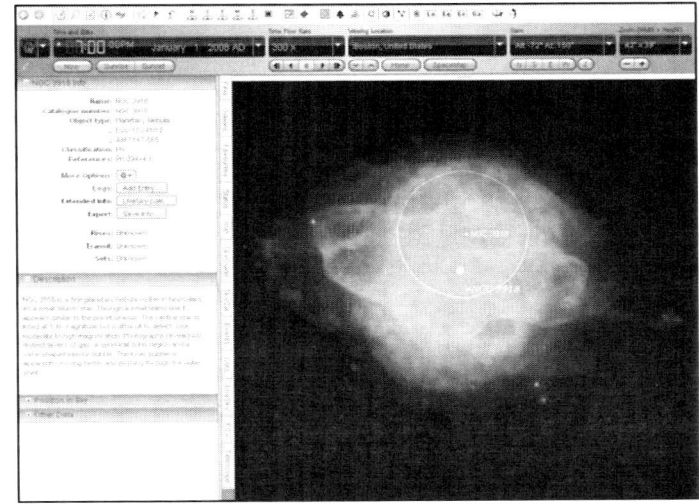

Look for NGC 3918 labeled within the boundaries of the Centaurus constellation as shown on your screen. If an object is not labeled on your screen, hovering your mouse pointer over the object will give you some identifying information. Once you locate what you are looking for, zoom in on the object using the **zoom controls** at the right of the **control panel**. Right click (control click for Mac) on the object and select **show info**, click the plus sign (gray arrow for Mac) to expand all layers, then complete the following table of information. If you have trouble finding NGC 3918, click the FIND tab on the left **side pane** and follow the optional procedure given previously.

Date, time:	
Azimuth, altitude:	
Distance from observer (give units):	
Apparent magnitude:	
Angular size (give units—degrees, arc minutes, etc.):	

What do you find most interesting? Indicate some additional info about this object:

NGC 3766 (Open Cluster):

NGC 3766 is an open cluster. Open clusters are groups of stars that have recently formed from the same interstellar cloud of gas and dust. They are relatively young in age and some can be seen with the naked eye and/or with binoculars.

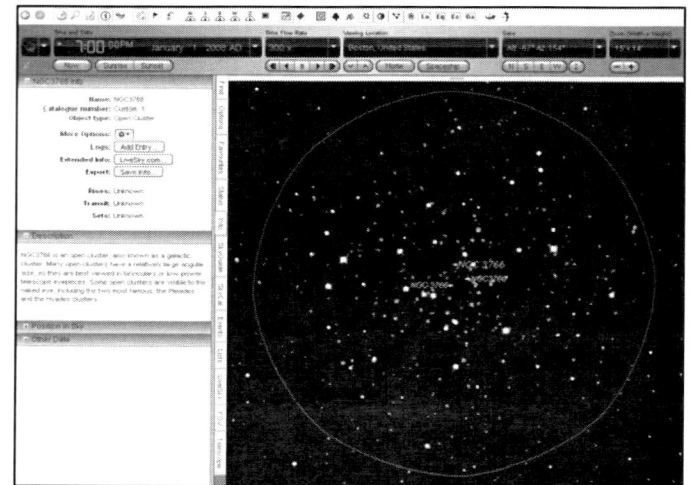

Look for NGC 3766 labeled within the boundaries of the Centaurus constellation as shown on your screen. If an object is not labeled on your screen, hovering your mouse pointer over the object will give you some identifying information. Once you locate what you are looking for, zoom in on the object using the **zoom controls** at the right of the **control panel**. Right click (control click for Mac) on the object and select **show info**, click the plus sign (gray arrow for Mac) to expand all layers, then complete the following table of information. If you have trouble finding NGC 3766, click the FIND tab on the left **side pane** and follow the optional procedure given previously.

Date, time:	
Azimuth, altitude:	
Distance from observer (give units):	
Apparent magnitude:	
Angular size (give units—degrees, arc minutes, etc.):	

What do you find most interesting? Indicate some additional info about this object:

Observation and Research Projects

Steve McMillan

DREXEL UNIVERSITY

Observation and Research Projects

Contents

CHARTING the HEAVENS

The Foundations of Astronomy

Observation & Research Projects

1. Go to a country location on a clear, dark night. Imagine patterns among the stars, and name the patterns yourself. Note (or better yet, draw) the locations of these stars with respect to trees or buildings in the foreground. Do this every week or so for a couple of months, and be sure to look at the same time every night. What happens?

2. Find the star Polaris, also known as the North Star, in the evening sky. Identify any separate pattern of stars in the same general vicinity of the sky. Wait several hours, at least until after midnight, and then locate Polaris again. Has Polaris moved? What has happened to the nearby pattern of stars? Why?

3. Hold your little finger out at arm's length. Can you cover the disk of the Moon? The Moon projects an angular size of 30′ (half a degree); your finger should more than cover it. How can you apply this fact in making sky measurements?

THE COPERNICAN REVOLUTION

The Birth of Modern Science

Observation & Research Projects

1. Look in an almanac for the date of opposition of one or all of these bright planets: Mars, Jupiter, and Saturn. At opposition, these planets are at their closest points to Earth and are at their largest and brightest in the night sky. Observe these planets. How long before opposition does each planet's retrograde motion begin? How long afterward does it end?

2. Draw an ellipse. (See Figure 2.15 in *Astronomy Today 7th Edition*.) You'll need two pins, a piece of string, and a pencil. Tie the string in a loop and place it around the pins. Place the pencil inside the loop and run it around the inside of the string, holding the loop taut. The two pins will be at the foci of the ellipse. What is the eccentricity of the ellipse you have drawn?

3. Use a small telescope to replicate Galileo's observations of Jupiter's four largest moons. Note the moons' brightnesses and their locations with respect to Jupiter. If you watch over a period of several nights, draw what you see; you'll notice that these moons change their positions as they orbit the giant planet. Check the charts given monthly in *Astronomy* or *Sky & Telescope* magazines to identify each moon you see.

CHAPTER

RADIATION

Information from the Cosmos

Observation & Research Projects

1. Locate the constellation Orion. Its two brightest stars are Betelgeuse and Rigel. Which of these is the hotter star? Which is cooler? How can you tell? Which of the other stars scattered across the night sky are hot, and which are cool?

2. Stand near (but not too near!) a train track or busy highway and wait for a train or traffic to pass by. Can you notice the Doppler effect in the pitch of the engine noise or whistle blowing? How does the sound frequency depend on (a) speed and (b) the train's or traffic's motion toward or away from you?

SPECTROSCOPY

The Inner Workings of Atoms

Observation & Research Projects

1. Find a spectrum of the Sun that also has a wavelength scale alongside. Figure 16.9 is a good example; however, you may want to enlarge it on a copying machine. Select various absorption lines and determine their wavelengths by interpolation. Now, try to identify the element that produced these lines. Use a reference of lines such as found in Moore's *A Multiplet Table of Astrophysical Interest*. Other references may be found in the astronomy, chemistry, or physics sections of your library. Work with the darkest lines before trying the fainter lines.

2. Use a handheld spectroscope, available through Learning Technologies, Inc. While in the shade, point the spectroscope at a white cloud or white piece of paper that is in direct sunlight. Look for the absorption lines in the Sun's spectrum. Note their wavelength from the scale inside the spectroscope. Compare your list with the Fraunhofer lines given in many physics, astronomy, or chemistry reference books.

CHAPTER

TELESCOPES

The Tools of Astronomy

Observation & Research Projects

1. Here's how to take some easy pictures of the night sky. You will need a location with a clear, dark sky; a 35-mm camera with a standard 50-mm lens, tripod, and cable release; a watch with a seconds display visible in the dark; and a roll of high-speed color film. Set your camera to the "bulb" setting for the exposure and attach the cable release so you can take a long exposure. Set the focus on infinity. Point the camera to a favored constellation, seen through your viewfinder, and take a 20–30-second exposure. Don't touch any part of the camera or hold on to the cable release during the exposure; minimize all vibrations. Keep a log of your shots. When finished, have the film developed in the standard way.

2. For some variation, vary your exposure times, use different films, take hours-long exposures for star trails, use different lenses such as wide-angle or telephoto, place the camera piggyback on a telescope that is tracking, and take exposures that are a few minutes long. Experiment and have fun!

THE SOLAR SYSTEM

An Introduction to
Comparative Planetology

Observation & Research Projects

1. You can begin to visualize the ecliptic—the plane of the planets' orbits—just by noticing the path of the Sun throughout the day and of the full Moon in the course of a single night. It helps if you watch from one spot, such as your backyard or a rooftop. It's also good to have a general notion of direction. (West is where the Sun sets!) You will see that the movements of the Sun and Moon are confined to a narrow pathway across our sky. The planets also travel along this path. The motion of the Sun, Moon, and planets is a two-dimensional reflection of the three-dimensional plane of our solar system.

2. Once you get a feeling for the whereabouts of the ecliptic, try locating the North Star. Knowing the direction to celestial north makes it easier to imagine the motion of the planets in the plane of the solar system. Don't worry about being too precise. Just get a sense of the ecliptic as a kind of merry-go-round of planets—that we on Earth also ride!

3. Go to your library or the Internet and find what planetary missions are in progress or are planned, other than those described in *Astronomy Today, 7th Edition*. Apart from the United States, what other countries have space agencies actively engaged in planetary exploration? What nonplanetary missions are in progress or planned?

CHAPTER

EARTH

Our Home in Space

Observation & Research Projects

1. Go to a sporting goods store and get a tide table; many stores near the ocean provide them for free. Choose a month and plot the height of one high and one low tide versus the day of the month. Now mark the dates when the primary phases of the Moon occur. How well does the phase of the Moon predict the tides?

2. Measure Earth's radius. You will need a friend or colleague (or another astronomy student with a project assignment!) who lives a few hundred kilometers due north or south of you. On the day of the first quarter Moon, right at sunset, you should both estimate, to within a tenth of a degree, the angular distance of the Moon above your southern horizon. Compare the angles you obtain; they should be different. Call this difference θ (the Greek letter theta). Determine the exact distance between your two locations using a map; call this d. Earth's radius can then be computed from the equation $r = 57.3 \, d / \theta$. Many details of how to do this experiment have been left for you to figure out. While you are at it, show where the formula comes from!

3. Go to a library or the Internet and read about global warming. How much carbon dioxide is produced each year by human activities? How does this compare with the total amount of carbon dioxide in Earth's atmosphere? What natural processes tend to reduce the level of atmospheric carbon dioxide? Do all scientists agree that global warming is an inevitable consequence of carbon dioxide production? What political initiatives are currently under way to address the problem?

THE MOON AND MERCURY

Scorched and Battered Worlds

Observation & Research Projects

1. Observe the Moon during an entire cycle of phases. When does the Moon rise, set, and appear highest in the sky at each major phase? What is the interval of time between each phase?

2. If you have binoculars, turn them on the Moon when it appears at twilight and when it appears high in the sky. Draw pictures of what you see. What differences do you notice in your two drawings? What color is the Moon seen near the horizon? What color is the Moon seen high in the sky? Why is there a difference?

3. Watch the Moon over a period of hours on a night when you can see one or more bright stars near it. Estimate how many Moon diameters it moves per hour, relative to the stars. Knowing the Moon is about 0.5° in diameter, how many degrees per hour does it move? What is your estimate of its orbital period?

4. Try to spot Mercury in the morning or evening twilight. (Hint: As seen from the Northern Hemisphere, the best evening apparitions of the planet take place in the spring, and the best morning apparitions take place in the fall.)

VENUS

Earth's Sister Planet

Observation & Research Projects

1. Is Venus in the morning or evening sky right now? Look for it every few days, over the course of several weeks. Draw a picture of the planet with respect to foreground trees or buildings. If you always observe at the same time every day, you may begin to notice that the planet is getting higher or lower in the sky.

2. Consult an almanac to determine the next time Venus will pass between Earth and Sun. How many days before and after this event can you glimpse the planet with the eye alone?

3. Consult the almanac again to find out the next time Venus will pass on the far side of the Sun from Earth. How many days before and after this event can you see the planet with the naked eye?

4. When Venus ornaments the predawn sky, try keeping track of the planet with your eye alone until it appears in a blue sky, after sunrise. As always, be careful not to look at the Sun!

5. Using a powerful pair of binoculars or a small telescope, examine Venus as it goes through its phases. Note the phase and the relative size of it. (You can compare its size to the field of view in a telescope; always use the same eyepiece for this.) Look at it every few days or once a week. Make a table of the shape of the phase, the size, and the relative brightness to the naked eye. After you have observed it through a significant change in phase, can you see the correlations between these three properties first recognized by Galileo?

CHAPTER

MARS

A Near Miss for Life?

Observation & Research Projects

1. Track the motion of the Red Planet in front of the stars for several months following its return to the predawn sky. (Consult an almanac to determine where Mars will be in the sky this year.) You will see that Mars moves rapidly in front of the stars, crossing many constellation boundaries.

2. Several months before opposition, Mars begins retrograde motion. Chart the planet's motion in front of the stars to determine when it stops moving eastward and begins moving toward the west.

3. Notice the increase in Mars's brightness as it approaches opposition. Why is it getting brighter? What other planets appear in the sky now? How do their brightnesses compare with that of Mars?

4. Look at Mars with as large a telescope as is available to you; binoculars will not be of use. Prepare ahead of time and find out the Martian season that will be occurring at the time of your observation, which hemisphere will be tilted in Earth's direction, and what longitude will be pointing toward Earth at the time of observation. This information can be obtained from many different computer almanacs. Sketch what you see. Look very carefully and take your time. Afterward, try to identify the various features you have seen with known objects on Mars.

CHAPTER

JUPITER
Giant of the Solar System

Observation & Research Projects

1. Are there any stars in the night sky that look as bright as Jupiter? What other differences do you notice between Jupiter and the stars?

2. Use binoculars to look at Jupiter. Be sure to hold them steady (try propping your arms up on the hood of a car, or sitting down and bracing them against your knees). Can you see any of Jupiter's four largest moons? If you come back the following evening, the moons' relative positions will have changed. Have some changed more than others?

3. Through a telescope, you should be able to see the red-and-tan cloud bands of Jupiter, and you can clearly see some moons. Do the moons orbit in the equatorial plane? Before observing, look up the positions of the Galilean moons in a current magazine such as *Astronomy* or *Sky & Telescope*. Identify each of the moons. Watch Io over a period of at least an hour or more. Can you see its motion? Do the same for Europa.

CHAPTER

SATURN
Spectacular Rings
and Mysterious Moons

Observation & Research Projects

1. Saturn moves more slowly among the stars than any other visible planet. How many degrees per year does it move? Look in an almanac to see where the planet is now. What constellation is it in now? Where will it be in one year?

2. Binoculars may not reveal the rings of Saturn, but most small telescopes will. Use a telescope to look at Saturn. Does Saturn appear flattened? Examine the rings. How are they tilted? Can you see a dark line in the rings? This is the Cassini Division. It once was thought to be a gap in the rings, but the *Voyager* spacecraft discovered that it is filled with tiny ringlets. Can you see the shadow of the rings on Saturn?

3. While looking at Saturn through a telescope, can you see any of its moons? They line up with the rings; Titan is often the farthest out, and always the brightest. How many moons can you see? Use an almanac to identify each one you find.

URANUS AND NEPTUNE

The Outer Worlds of the Solar System

Observation & Research Projects

1. The major astronomy magazines *Sky & Telescope* and *Astronomy* print charts showing the whereabouts of the planets in their January issues. Consult one of these charts and locate Neptune and Uranus in the sky. Uranus may be visible to the naked eye, but binoculars make the search much easier. (Hint: Uranus shines more steadily than the background stars.) With the eye alone, can you detect a color to Uranus? Through binoculars?

2. The search for Neptune requires a much more determined effort! A telescope is best, but high-powered binoculars mounted on a steady support will do. If you can see both planets through a telescope—and they will remain close together on the sky for the rest of this century—contrast their colors. Which planet appears bluer? Through a telescope, does Uranus show a disk? Can you see that Neptune shows a disk, or does it look more like a point of light?

SOLAR SYSTEM DEBRIS

Keys to Our Origin

Observation & Research Projects

1. The only way to tell an asteroid from a star is to watch it over several nights. You can detect its movement in front of the star background. The astronomy magazines *Sky & Telescope* and *Astronomy* often publish charts for especially prominent asteroids. Look for the asteroids Ceres, Pallas, or Vesta. They are the brightest asteroids. Use the chart to locate the appropriate star field. Aim binoculars at that location in the sky; you may be able to pick out the asteroid from its location in the chart. If you can't, make a rough drawing of the entire field. Come back a night or two later, and look again. The "star" that has moved is the asteroid.

2. Although a spectacular naked-eye comet comes along only about once a decade, fainter comets can be seen with binoculars and telescopes in the course of every year. *Sky & Telescope* often runs a "Comet Digest" column announcing the whereabouts of comets. In addition, a comprehensive list of periodic comets expected to return in a given year can be found in *Guy Ottewell's Astronomical Calendar*, which contains a wealth of other sky information as well, including monthly star charts. At the time of this writing, it costs $26.95 per year and can be purchased at www.universalworkshop.com.

3. There are a number of major meteor showers every year, but if you plan to watch one, be sure to notice the phase of the Moon. Bright moonlight or city lights can obliterate a meteor shower. A common misconception about meteor watching is that most meteors are seen in the direction of the shower's radiant point. It's true that if you trace the paths of the meteors backward in the sky, they all can be seen to come from the radiant. But most meteors don't become visible until they are 20° or 30° from the radiant. Meteors can appear in all parts of the sky! Just relax and let your eyes rove among the stars. You will generally see many more meteors in the hours before dawn than in the hours after sunset. Why do you suppose meteors have different brightnesses? Can you detect their variety of colors? Watch for meteors that appear to "explode" as they fall, and vapor trails that linger after the meteor itself has disappeared.

CHAPTER

THE FORMATION OF PLANETARY SYSTEMS

The Solar System and Beyond

No Observation & Research Projects for Chapter 15

THE SUN

Our Parent Star

Observation & Research Projects

The projects given here require a special solar filter. Such filters are easily purchased from various sources.

NEVER LOOK DIRECTLY AT THE SUN WITHOUT A FILTER!

1. An appropriately filtered telescope will easily show you sunspots. Count the number of sunspots you see on the Sun's surface. Notice that sunspots often come in pairs or groups. Come back and look again a few days later and you'll see that the Sun's rotation has caused spots to move, and the spots themselves have changed. If a sufficiently large sunspot (or, more likely, sunspot group) is seen, continue to watch it as the Sun rotates. It will be out of view for about two weeks. Can you determine the rotation of the Sun from these observations?

2. Solar granulation is not too hard to see. The atmosphere of Earth is most stable in the morning hours. Observe the Sun on a cool morning, one or two hours after it has risen. Use high magnification and look initially at the middle of the Sun's disk. Can you see changes in the granulation pattern? They are there, but they are not always obvious or easy to see.

3. View some solar prominences and flares. Hydrogen-alpha (H) filters commercially available for small telescopes are quite expensive, but many science departments will have one. You can often see prominences and flares even during times of sunspot minimum. You are actually viewing the chromosphere rather than the photosphere, so the Sun looks quite different from its normal appearance.

THE STARS

Giants, Dwarfs, and the
Main Sequence

Observation & Research Projects

1. Every winter, you can find an astronomy lesson in the evening sky. The Winter Circle is an asterism—or pattern of stars—made up of six bright stars in five different constellations: Sirius, Rigel, Betelgeuse, Aldebaran, Capella, and Procyon. These stars span nearly the entire range of colors (and therefore temperatures) possible for normal stars. Rigel is a B-type star, Sirius is an A-type, Procyon is an F-type star, Capella is a G-type star, Aldebaran a K-type star, and Betelgeuse is an M-type star. The color differences of these stars are easy to see. Why do you suppose there is no O-type star in the Winter Circle?

2. In the winter sky, you'll find the red supergiant Betelgeuse in the constellation Orion. It's easy to see because it's one of the brightest stars visible in our night sky. Betelgeuse is a variable star with a period of about 6.5 years. Its brightness changes as it expands and contracts. At maximum size, Betelgeuse fills a volume of space that would extend from the Sun to beyond the orbit of Jupiter. Betelgeuse is thought to be about 10 to 15 times more massive than our Sun, and probably between four and 10 million years old. A similar star can be found shining prominently in midsummer. This is the red supergiant Antares in the constellation Scorpius. Depending on the time of year, can you find one of these stars? Why are they red?

THE INTERSTELLAR MEDIUM

Gas and Dust among the Stars

Observation & Research Projects

1. The constellation Orion, the Hunter, is prominent in the evening sky of winter. Its most noticeable feature is a short, straight row of three medium-bright stars: the famous belt of Orion. A line of stars extends from the easternmost star of the belt, toward the south. This line represents Orion's sword. Towards the bottom of the sword is the sky's most famous emission nebula, M42, the Orion Nebula. Observe the Orion Nebula with your eye, with binoculars, and with a telescope. What is its color? How can you account for this? With the telescope, try to find the Trapezium, a grouping of four stars in the center of M42. These are hot, young stars; their energy causes the Orion Nebula to glow.

2. Observe the Milky Way on a dark, very clear night. Is it a continuous band of light across the sky or is it mottled? The parts of the Milky Way that appear missing are actually dark dust clouds that are relatively near the Sun. Identify the constellations in which you see these clouds. Make a sketch and compare with a star atlas. Find other small clouds in the atlas and try to find them with your eye or with binoculars.

STAR FORMATION

A Traumatic Birth

Observation & Research Projects

1. The Trifid Nebula, otherwise known as M20, is a place where new stars are forming. It has been called a "dark night revelation, even in modest apertures." An 8- to 10-inch telescope is needed to see the triple-lobed structure of the nebula. Ordinary binoculars reveal the Trifid as a hazy patch located in the constellation Sagittarius. This nebula is set against the richest part of the Milky Way, the edgewise projection of our own Galaxy around the sky. It is one of many wonders in this region of the heavens. What are the dark lanes in M20? Why are other parts of the nebula bright? There have been reports of large-scale changes occurring in this nebula in the last century and a half. The reports are based on old drawings, which show M20 looking slightly different from how it appears today. Do you think it possible for a cloud in space to undergo a change in appearance on a time scale of years, decades, or centuries?

2. Summer is a good time to search with binoculars for open-star clusters. Open clusters are generally found in the plane of the Galaxy. If you can see the hazy band of the Milky Way arcing across your night sky—in other words, if you are far from city lights and looking at an appropriate time of night and year—you can simply sweep with your binoculars along the Milky Way. Numerous "clumps" of stars will pop into view. Many will turn out to be open-star clusters.

3. Globular star clusters are more difficult to find. They are intrinsically larger, but they are also much farther away and therefore appear smaller in the sky. The most famous globular cluster visible from the Northern Hemisphere is M13 in the constellation Hercules, visible on spring and summer evenings. This cluster contains half a million or so of the Galaxy's most ancient stars. It may be glimpsed in binoculars as a little ball of light, located about one-third of the way from the star Eta to the star Zeta in the Keystone asterism of the constellation Hercules. Telescopes reveal this cluster as a magnificent, symmetrical grouping of stars.

STELLAR EVOLUTION

The Life and Death of a Star

Observation & Research Projects

1. Can you find the Hyades cluster? It lies about 46 pc away in the constellation Taurus, making up the "face" of the bull. It appears to surround the very bright star Aldebaran, the Bull's eye, which makes it easy to locate in the sky. Aldebaran is a red giant, probably on the asymptotic-giant branch of its evolution. Despite appearances, it is not part of the Hyades cluster. In fact it lies only about half as far away—some 20 pc from Earth.

2. Now look for the Double Cluster in Perseus, h and chi Persei. These two young clusters probably formed together, and now move together through space. They lie about 2500 pc away, and are barely visible to the naked eye just east of the "W" of Cassiopeia.

3. Find a library that has the *Astrophysical Journal*. Find an article from the late 1950s and 1960s that gives the photometry of a star cluster like the Pleiades or Hyades. Plot a color–magnitude diagram (V vs. B–V; see *Astronomy Today, 7th edition*, Section 17.5). Determine the V magnitude of the main-sequence turnoff, and hence estimate the age of the cluster. Compare your age with that given in the article.

STELLAR EXPLOSIONS

Novae, Supernovae, and the
Formation of the Elements

Observation & Research Projects

1. In 1758, the French comet hunter, Charles Messier discovered the sky's most legendary supernova remnant, now called M1, or the Crab Nebula. It is located northwest of Zeta Tauri, the star that marks the southern tip of the horns of Taurus the Bull. Try to find it—an 8-inch telescope reveals the Crab's oval shape, but it will appear faint; a 10-inch or larger telescope reveals some of its famous filamentary structure.

2. In the *Handbook of Chemistry and Physics*, available in the library reference section, look up the table of isotopes. Pick one or more isotopes and follow their decay into a final stable isotope. For example, choose cobalt-59, formed in the s-process. Note how the isotope decays, what is emitted, and the half-life of the decays. Try this exercise for uranium-235, uranium-238, and plutonium-239.

CHAPTER

NEUTRON STARS AND BLACK HOLES

Strange States of Matter

Observation & Research Projects

1. Many amateur astronomers enjoy turning their telescopes on the magnitude-9 companion to Cygnus X-1, the sky's most famous black-hole candidate. Because none of us can see in X-rays, no sign of anything unusual can be seen. Still, it's fun to gaze toward this region of the heavens and contemplate Cygnus X-1's powerful energy emission and strange properties. Even without a telescope, it is easy to locate the region of the heavens where Cygnus X-1 resides. The constellation Cygnus contains a recognizable star pattern, or asterism, in the shape of a large cross. This asterism is called the Northern Cross. The star in the center of the crossbar is called Sadr. The star at the bottom of the cross is called Albireo. Approximately midway along an imaginary line between Sadr and Albireo lies the star Eta Cygni. Cygnus X-1 is located slightly less than 0.5° from this star. With or without a telescope, sketch what you see.

2. Set up a demonstration of the densities of various astronomical objects—an interstellar cloud, a star, a terrestrial planet, a white dwarf, and a neutron star. Select a common object that is easily held in your hand, something that would be familiar to anyone—an apple, for example. For the lowest densities, calculate how large a volume would contain the object's equivalent mass. For high densities, calculate how many of the objects would have to be fit into a standard volume, such as 1 cm^3. This volume is better for this project than 1 m^3 because most people do not appreciate how large a volume 1 m^3 is. Present your demonstration to your class or to some other group of students. Tell them about each astronomical object and how it comes by its density.

233

THE MILKY WAY GALAXY

A Spiral in Space

Observation & Research Project

1. If you are far from city lights, look for a hazy band of light arching across the sky. This is our edgewise view of the Milky Way Galaxy. The Galactic center is located in the direction of the constellation Sagittarius, highest in the sky during the summer, but visible from spring through fall. Look at the band making up the Milky Way and notice the dark regions; these are relatively nearby dust clouds. Sketch what you see. Look for faint fuzzy spots in the Milky Way and note their positions in your sketch. Draw in the major constellations for reference. Compare your sketch with a map of the Milky Way in a star atlas. Did you discover most of the dust clouds? Can you identify the faint fuzzy spots?

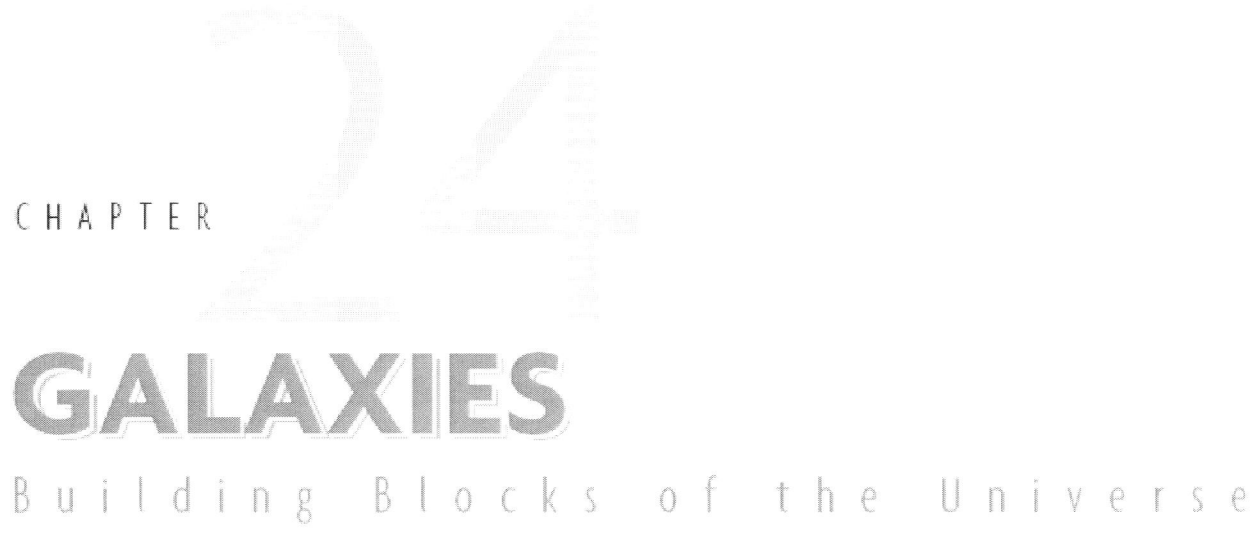

GALAXIES

Building Blocks of the Universe

Observation & Research Projects

1. Look for a copy of the *Atlas of Peculiar Galaxies* by Halton Arp. It is available in book form or on laser disk. Search for examples of interacting galaxies of various types: (1) tidal interactions, (2) starburst galaxies, (3) collisions between two spirals, and (4) collisions between a spiral and an elliptical. For (1) look for galactic material pulled away from a galaxy by a neighboring galaxy. Is the latter galaxy also tidally distorted? In (2) the surest signs of starburst activity are bright knots of star formation. In what type(s) of galaxies do you find starburst activity? For (3) and (4) how do collisions differ depending on the types of galaxies involved? What typically happens to a spiral galaxy after a near miss or collision? Do ellipticals suffer the same fate?

2. Look for the Virgo Cluster of galaxies. An 8-inch telescope is the perfect size for this project, although a smaller telescope will also work. The constellation Virgo is visible from the United States during much of fall, winter, and spring. To locate the center of the cluster, first find the constellation Leo. The eastern part of Leo is composed of a distinct triangle of stars, Denebola (β Leo), Chort (θ Leo), and Zosma (δ Leo). Go from Chort to Denebola in a straight line east, continue on the same distance as between the two stars and you will be approximately at the center of the Virgo Cluster. Look for the following Messier objects that make up some of the brightest galaxies in the cluster: M49, M58, M59, M60, M84, M86, M87 (a giant elliptical thought to have a massive black hole at its center), M89, and M90. Examine each galaxy for unusual features; some have very bright nuclei.

GALAXIES AND DARK MATTER

The Large-Scale Structure of the Cosmos

Observation & Research Projects

Here are three observational projects that are increasingly challenging.

1. In the previous chapter you were given directions for finding the Virgo Cluster of galaxies. M87, in the central part of this cluster, is the nearest core–halo radio galaxy. M87 has coordinates R.A. 12h 30.8m, Dec. + 12° 24'. At magnitude 8.6, it should not be difficult to find in an 8-inch telescope. Its distance is roughly 20 Mpc. Describe its nucleus; compare what you see with other nearby ellipticals in the Virgo Cluster.

2. NGC 4151 is the brightest Seyfert galaxy. Its coordinates are R.A. 12h 10.4m, Dec. + 39° 24', and it can be found below the Big Dipper in Canes Venatici. At magnitude 10–12 (it is variable), it should be visible in an 8-inch telescope, but it will be challenging to find. Its distance is 13.5 Mpc. As in the case of M87, describe its nucleus and compare with what you have seen for other galaxies.

3. 3C 273 is the nearest and brightest quasar. However, that does not mean it will be easy to find and see! Its coordinates are RA 12h 29.2m, Dec. + 2° 03'. It is located in the southern part of the Virgo Cluster but is not associated with it. At magnitude 12–13 (again, it is variable), it may require a 10- or 12-inch telescope to see, but try it first with an 8-inch telescope. It should appear as a very faint star. The significance of seeing this object is that it is 640 Mpc distant. The light you are seeing left this object over 2 billion years ago! 3C 273 is the most distant object observable with a small telescope.

If you can find the three objects listed here, you have started to become an accomplished observer!

CHAPTER

COSMOLOGY

The Big Bang and the Fate
of the Universe

Observation & Research Projects

1. Make a model of a two-dimensional universe and examine Hubble's law on it. Find a balloon that will expand into a nice large sphere. Blow it up about halfway and mark dots all over its surface; the dots will represent galaxies. Choose one dot as your home galaxy. Using a measuring tape, measure the distances to various other galaxies, numbering the dots so you will not confuse them later. Now blow the balloon up to full size and measure the distances again, and find the new distances to each dot. Calculate the change in the distances for each galaxy; this is a measure of their velocity (change in position/change in time; the time is the same for all and is arbitrary). Plot their velocities versus their new distances as in Figure 17.6. Do you get a straight-line correlation, i.e., a "Hubble" law? Does it matter which dot you choose as home? Demonstrate this to your class.

2. Write a paper on the philosophical differences between living in an open, closed, or flat universe. Are there aspects of any of these three possibilities that are hard to accept? It is quite possible that astronomers may determine within your lifetime which is correct. Do you have a preference?

3. Go to your library or the Internet and read about the steady-state universe, which enjoyed some measure of popularity in the 1950s and 1960s. How does it differ from the standard Big Bang model? Why do you think the steady-state model is not widely accepted today?

THE EARLY UNIVERSE

Toward the Beginning of Time

Observation & Research Projects

1. Read the book *The First Three Minutes* by Steven Weinberg .It is fairly nonmathematical in its presentation. What new results are presented in this chapter that were not known by Weinberg when he wrote the first edition in 1977? How much progress has been made in understanding the very earliest epochs since that time?

2. Write a paper on the cosmological constant. Is there presently any observational evidence favoring a nonzero value for this constant? What are the main theoretical reasons for including a cosmological constant in theories of the universe?

LIFE IN THE UNIVERSE

Are We Alone?

Observation & Research Projects

1. Some people suggest that if extraterrestrial life is discovered, it will have a profound effect on people. Interview as many people as you can and ask the following two questions:

    (1) Do you believe that extraterrestrial life exists? (2) Why? From your results, try to decide whether there will be a profound effect on people if extraterrestrial life is discovered.

2. Conduct another poll, or do it at the same time as the first one. Ask the following question: What one question would you like to ask an extraterrestrial life form in a radio communication? How many responses do you receive that indicate the person is very "Earth-centered" in thinking? How many responses suggest a lack of understanding of how alien an extraterrestrial life form might be? Is your conclusion from the first project different or changed in any way?

3. The Drake equation should be able to "predict" at least one civilization in our Galaxy: us. Try changing the values of various factors so that you end up with at least one. What do these various combinations of factors imply about how life arises and develops? Are there some combinations that just don't make any sense?